P Barber

IUPAB Biophysics Series
sponsored by
The International Union of Pure and Applied Biophysics

Basic principles of membrane transport

IUPAB Biophysics Series
sponsored by
The International Union of Pure and Applied Biophysics
Editors:
Franklin Hutchinson
Yale University
Watson Fuller
University of Keele
Lorin J. Mullins
University of Maryland

1 Walter Harm: *Biological effects of ultraviolet radiation*
2 Stanley G. Schultz: *Basic principles of membrane transport*
3 E. G. Richards and S. D. Dover: *An introduction to the physical properties of large molecules in solution*
4 Roderick K. Clayton: *Photosynthesis: physical mechanisms and chemical patterns*

Basic principles of membrane transport

STANLEY G. SCHULTZ

Professor and Chairman, Department of Physiology
University of Texas Medical School
Houston, Texas

CAMBRIDGE UNIVERSITY PRESS

CAMBRIDGE

LONDON NEW YORK NEW ROCHELLE

MELBOURNE SYDNEY

Published by the Press Syndicate of the University of Cambridge
The Pitt Building, Trumpington Street, Cambridge CB2 1RP
32 East 57th Street, New York, NY 10022, USA
296 Beaconsfield Parade, Middle Park, Melbourne 3206, Australia

First published 1980

Printed in the United States of America
Typeset by Automated Composition Service, Inc., Lancaster, Pennsylvania
Printed and bound by Vail-Ballou Press, Inc., Binghamton, New York

Library of Congress Cataloging in Publication Data

Schultz, Stanley G.

Basic principles of membrane transport.

(IUPAB biophysics series; 2)
1. Biological transport. I. Title. II. Series: International Union for Pure and Applied Biophysics. IUPAB biophysics series; 2.
QH509.S38 574.8′75 79-54015
ISBN 0 521 22992 8 hard covers
ISBN 0 521 29762 1 paperback

CONTENTS

FOREWORD

The origins of this series were a number of discussions in the Education Committee and in the Council of the International Union of Pure and Applied Biophysics (IUPAB). The subject of the discussions was the writing of a textbook in biophysics; the driving force behind the talks was Professor Aharon Katchalsky, first while he was president of the Union, and later as the honorary vice-president.

As discussions progressed, the concept of a unified text was gradually replaced by that of a series of short inexpensive volumes, each devoted to a single topic. It was felt that this format would be more flexible and more suitable in light of the rapid advances in many areas of biophysics at present. Instructors can use the volumes in various combinations according to the needs of their courses; new volumes can be issued as new fields become important and as current texts become obsolete.

The International Union of Pure and Applied Biophysics was motivated to participate in the publication of such a series for two reasons. First, the Union is in a position to give advice on the need for texts in various areas. Second, and even more important, it can help in the search for authors who have both the specific scientific background and the breadth of vision needed to organize the knowledge in their fields in a useful and lasting way.

The texts are designed for students in the last years of the standard university curriculum and for Ph.D. and M.D. candidates taking advanced courses. They should also provide a suitable introduction for someone about to begin research in a particular field of biophysics. The Union is pleased to collaborate with the Cambridge University Press in making these texts available to students and scientists throughout the world.

Franklin Hutchinson, Yale University
Watson Fuller, University of Keele
Lorin J. Mullins, University of Maryland
Editors

FOREWORD

PREFACE

My initial reaction upon being invited by the Editorial Committee of the International Union of Pure and Applied Biophysics to write a monograph on principles of membrane transport was a mixture of flattery and terror. I was naturally flattered by the invitation but terrified by the seeming enormity of the task and the effect that acquiescence would have on my already overburdened schedule. When this initial reaction gave way to calm objectivity, I asked myself: With so many books appearing annually on the subject of membrane transport, is there a justifiable need for yet another?

My answer to this question, which was confirmed by many of my colleagues, was that there still is a need for the type of "primer" I would have desperately longed for in 1959 when, after completing a Residency in Internal Medicine, I joined the Biophysical Laboratories of the Harvard Medical School as an initiate in the field of membrane transport; my contacts with undergraduates, graduate students, postdoctoral fellows, and junior scientists during the past two decades have reinforced this feeling.

In short, this monograph is intended for the "initiate." Its goal is to introduce some of the basic principles that govern solute and water transport across membranes and to emphasize the foundations and reasoning that underlie these principles in a relatively brief and readable form. It is not intended to supplant the more advanced and comprehensive treatments that can be found in a number of books, chapters, and review articles, but rather to provide the readers with an appreciation of basic concepts and approaches that will ease their way into the more advanced literature. In order to achieve this goal, I have attempted to develop this monograph from "first principles" assuming only that the reader has had some previous exposure to elementary differential and integral calculus, chemistry, and physics. All final expressions are derived, inasmuch as possible, step by step, and particular care has been taken to see to it that a consistent set of conventions is carefully defined and adhered to throughout. There are a number of basic points that are repeated throughout this monograph; this may irritate the more advanced reader but, at the same time, could be valuable to the beginner. In any event, this introductory text is designed to teach and I am convinced that repetition is an essential ingredient of pedagogy.

If I have succeeded in achieving my intended goal, a large measure of the success is owed to my teachers, colleagues, and students.

I am particularly indebted to:

Professor Arthur K. Solomon, who opened the door to my scientific career and who permitted (urged) me to teach the first offering of a course entitled "Bioenergetics and Membrane Transport" (Biology 119) at Harvard College in 1967. This was a memorable experience which forced me to come to grips with theoretical principles from the vantage point of a teacher rather than a user. Much of the organization and many of the developments of this text are based on my lecture notes for that course.

The late Professor Peter F. Curran, a long-time combination teacher-collaborator-friend, who was exceptionally capable of employing theory to provide a logical, coherent, and concise "picture" of experimental data.

And, the late Professor Aharon Katchalsky, a profound thinker, a gifted educator, and a constant source of inspiration; one cannot lavish too much praise upon his personal and intellectual attributes.

The tragic and untimely deaths of Peter Curran and Aharon Katchalsky, within a brief span of time, came as terrible blows. I am privileged to have shared their friendship and collegiality, and owe them both a great deal.

In addition, I am grateful to Drs. Alvin Essig, Raymond Frizzell, and Stephen Thompson for their careful reading of this manuscript and their corrections, cogent suggestions, and critical comments.

I wish to thank my secretary, Susan Frizzel, who nurtured this effort from first draft to final product and am eternally indebted to her dog, Fagin, for not having completely devoured Chapter 4 during an unguarded moment. (I reject my son's suggestion that he found it "too dry"!)

Finally, I owe a special brand of gratitude to my wife, Harriet, and my children, Jeffery and Kenneth, who have always provided me with the warmth, encouragement, and peace of mind to enable me to "do my thing." My debt to them and my parents can never be repaid.

S.G.S.

1 Some basic principles of thermodynamics: The relations between flows and forces

The formal description of transport across membranes rests firmly on the unassailable citadel of thermodynamics, that branch of science concerned with the changes that accompany the transition of a system (a defined portion of the universe selected for study) from one set of conditions (state) to another.[1] In this chapter we will develop the basic principles that deal with the transitions of a system from one state (initial) to another (final), which result from the transfer of matter within the system.[2]

1.1. Properties of state

Any macroscopic system at equilibrium can be described completely by an *equation of state*, which relates a set of parameters or properties that are functions only of the current state of the system and are independent of its past history. It follows that any change in a *property of state* is uniquely determined by the initial and final conditions of the system and is independent of the path (or mechanism) of transition between these two states.

A *property of state* can be formally defined as follows. If Z is a function of other properties of state, for example, $Z = f(U, V, W, \ldots)$ and if

$$dZ = L \cdot dU + M \cdot dV + N \cdot dW + \cdots$$

where $L = (\partial Z/\partial U)_{V,W,\ldots}$, $M = (\partial Z/\partial V)_{U,W,\ldots}$, $N = (\partial Z/\partial W)_{U,V,\ldots}$, and so on, then dZ is an *exact* (*or total*) *differential* of $f(U, V, W, \ldots)$ and Z is a property of state.[3] The reason, of course, is that if these relations hold, then $L \cdot dU$, $M \cdot dV$, $N \cdot dW$, and so on, represent the *individual contributions* of the changes in $U, V, W, \ldots$ to the *total change* in Z and, since addition is a commutative operation, the total change in Z does not depend upon the order or sequence of the changes in $U, V, W, \ldots$. For example, referring to Figure 1.1, if a system undergoes a change from state A to state C as a result of changes in properties U and V, the change in Z will be the same if the change in U preceded the change in V (pathway 1) or if both U and V change simultaneously (pathway 2), that is,

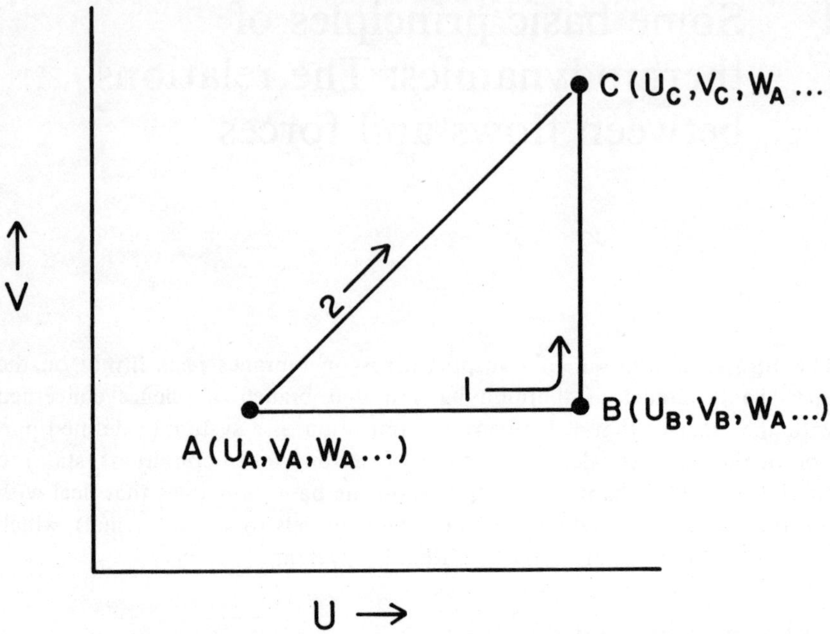

Figure 1.1. $U, V, W, \ldots$ are properties of state. If a system undergoes a transition from an initial state A to a final state C as a result of changes in U and V, the properties of the final state are independent of the path.

$$\Delta Z = Z_C - Z_A = f(U_C, V_C, W_A, \ldots) - f(U_A, V_A, W_A, \ldots)$$
$$= [(Z_B - Z_A) + (Z_C - Z_B)]$$

Finally, the fact that dZ is an exact differential means that there will be no change in Z if the system is displaced from state A and then restored to state A regardless of the paths (mechanisms and durations) of displacement and restoration (e.g., referring to Figure 1.1, A to B and back, A to C and back, A to B to C and back, A to C to B and back); that is, the closed integral, $\oint dZ = 0$ (Green's theorem).

The state properties of a homogeneous system can be divided into two categories, namely, *intensive properties* and *extensive properties.* Intensive properties are those characteristic of the total system, as well as any arbitrarily selected part of the system, and they are nonadditive; for example, pressure (P), temperature (T), concentration (c), and electrical potential (ψ); these properties reflect the "intensity" of the system. The extensive properties of a homogeneous system are measures of the size or content of the system and are additive; for example, volume (V), number of moles (n). Obviously, the total volume of a system is the sum of the volumes of all sub-

Table 1.1. *Conjugate properties of state*

Intensive property	Extensive property
Pressure (P)	Volume (V)
Electrical potential (ψ)	Charge (e)
Temperature (T)	Entropy (S)
Chemical potential (μ)	Moles (n)

divisions (parts) of the system whereas the temperature of these subdivisions is the same. Further, it is obvious that the ratio of any two extensive properties is an intensive property (e.g., $n/V = c$), and it can be shown that any intensive property can be expressed as a function of all other intensive properties of a system (e.g., the perfect gas law $PV = nRT$ can be written $P = cRT$).

Every intensive property is paired with a *conjugate* extensive property and a difference in an intensive property is *a* driving force for the flow or displacement of its conjugate extensive property. The conjugate properties of state with which we will be concerned are listed in Table 1.1. Clearly, a difference in pressure (ΔP) is a driving force for the displacement of volume (V); a difference in electrical potential ($\Delta\psi$) is a driving force for the flow of charge (e); as we will see, a difference in temperature (ΔT) is a driving force for the displacement of entropy (S); and for the moment let us define a driving force for the flow of uncharged matter (n) as a difference in chemical potential ($\Delta\mu$). An explicit expression for the chemical potential will be derived below. Further, for reasons that will become evident, we refer to a difference in an intensive property as the *conjugate driving force* for the flow of its paired (conjugate) extensive property. Finally, it should be noted that the product of an intensive property and its conjugate extensive property has units of energy or work. Thus, $P\,dV$ is the work that must be performed to displace an amount of volume (dV) against a pressure (P); $\psi\,de$ is the work that must be performed to bring an amount of charge (de) from a region of zero electrical potential (ground state) to a region where the electrical potential is ψ; and as we will see, $\mu_i\,dn_i$ is the work that must be performed to transfer a number of moles of an uncharged substance (dn_i) from a region where the chemical potential of i is zero to one where it is μ_i.

1.2. The first and second laws of thermodynamics

The entire edifice of thermodynamics is founded on two laws or postulates that simply summarize a vast amount of human experience. The *first law*, often referred to as the law of conservation of energy, states that the total internal energy of a system (E) can only change as a result of the gain or loss of heat (Q) and/or performance of work (W) on or by the system;

that is,

$$dE = dQ - dW \tag{1.1}$$

where dQ is positive when heat is gained by the system and dW is positive when work is performed by the system on its surroundings. E is an extensive property of state determined by the total potential and kinetic energy of the components within the system, and dE is an exact differential of an equation of state. However, Q and W are not properties of state inasmuch as the amount of heat gained and work performed by a system when it changes from one state to another depends upon the nature (pathway, mechanism) of the transition.

For a system that is capable of exchanging matter with its surroundings, equation (1.1) can be written more explicitly as follows:

$$dE = dQ - P\,dV + \psi\,de + \sum_i \mu_i\,dn_i \tag{1.2}$$

where $P\,dV$ is positive when the system performs work by expanding (dV) against an external pressure P; $\psi\,de$ is positive when an amount of charge de is transferred into a system whose electrical potential is ψ; and $\mu_i\,dn_i$ is positive when an amount (moles) of matter dn_i is transferred into a system where the chemical potential of i is μ_i (see note 9).

The first law is a "bookkeeper's delight"; it states that a change in the internal energy of a system must be the sum of the exchange of heat with its surroundings and the work performed on or by its surroundings. Since the heat *gained* or *lost* by the system must be equal to the heat *lost* or *gained* by the surroundings, and since the energy lost by a system as a result of the performance of work on its surroundings must be equal to the energy gained by the surroundings, the internal energy of the universe is conserved; E is simply transferred from one system to another but the total remains constant. However, the first law provides no way of distinguishing between spontaneous or natural processes and processes that can never take place spontaneously; according to this law, all processes are equally possible providing they obey equation (1.1).

The *second law* of thermodynamics is the culmination of many efforts to describe the fact that all natural or spontaneous processes take place only in the direction toward equilibrium and come to a halt when this time-independent state is reached! Work must be performed on a system to displace it from equilibrium; such a displacement cannot occur spontaneously and, for this reason, spontaneous processes are referred to as "irreversible." This monumental achievement was accomplished by introducing the concepts of reversible processes, namely, processes that take place infinitely slowly so that the transition from one state to another may be treated as if the system passes through an infinite series of equilibrium states. For such an idealized process, it can be shown that, although dQ *is not* an exact differential, the ratio (dQ/T) *is* an exact differential. Thus, a new property of state emerges, the entropy (S), which for a reversible process is defined by the relation

$$dS = dQ/T \tag{1.3}$$

Since S is a property of state, a change in S is dependent only on the initial and final states of the system and is independent of the intervening path. Thus, for a given set of initial and final conditions, dS is the same if the transition between these states is the result of a spontaneous (irreversible) process or an idealized reversible process. However, for a given transition, the heat lost to the surroundings is greater (or the amount of heat taken up by the system from its surroundings is less) when the process is irreversible than when it is reversible.[4] Thus, for an irreversible process

$$dS > dQ/T \tag{1.4}$$

Stated another way, if the same change in state is carried out reversibly and irreversibly

$$dS = (dQ/T)_{rev} > (dQ/T)_{irrev}$$

That is, the heat that is actually absorbed in any real process is less than what would have been absorbed had the process been reversible. The difference between $(dQ/T)_{rev}$ and $(dQ/T)_{irrev}$ was referred to by Clausius as "the uncompensated heat."

Equations (1.3) and (1.4) for reversible and irreversible processes, respectively, comprise a formal statement of the second law of thermodynamics.

The direction of a spontaneous or irreversible process must satisfy the inequality given in equation (1.4). Since S is an extensive property of state, dS can be divided into two parts, that is,

$$dS = d_eS + d_iS$$

where d_eS represents the change in entropy of a system caused by exchange of heat with the surroundings and d_iS represents the internal production or creation of entropy (the "uncompensated heat"). For an irreversible process $d_eS + d_iS > dQ/T$, where $d_eS = dQ/T$, so that $d_iS > 0$; for a reversible process $d_eS + d_iS = dQ/T$, so that $d_iS = 0$. Thus, an irreversible process can be defined as one that leads to an internal production of entropy that cannot be accounted for by an exchange of heat with the surroundings. Further, it should be noted that for an irreversible process, d_eS and dS can be positive or negative depending upon whether heat is gained or lost by the system, but d_iS is always positive.

Finally, since E is a property of state, dE is independent of path and for a given change in state is the same whether the transition is the result of a reversible or an irreversible process. Thus, we can substitute $T\,dS$ for dQ in equation (1-2) and obtain

$$dE = T\,dS - P\,dV + \psi\,de + \sum \mu_i\,dn_i \tag{1.5}$$

From this equation, it is clear that S is the extensive conjugate of T in the displacement or flow of thermal energy.[5]

Equation (1.5) is often referred to as the "Gibbs equation," and in its most general form it includes changes in all possible extensive properties and equates dE to the sum of the products of intensive properties and changes in their conjugate extensive properties, that is,

$$dE = \sum_i [(\text{Intensive property})_i \times (\text{Change in conjugate extensive property})_i]$$

1.3. The Gibbs free energy and the concept of "useful work"

Although the entire fabric of classical thermodynamics can be woven out of the first and second laws without introducing any additional properties of state, equation (1.5) does not provide the most convenient or illuminating framework for the thermodynamic description of processes that take place under conditions of constant temperature and pressure. To remedy this shortcoming, Gibbs introduced the free-energy function G, which is defined as

$$G = E + PV - TS \tag{1.6}$$

Since E, P, V, T, and S are properties of state, G must also be a property of state, so that dG is determined entirely by the initial and final conditions of a system and is independent of the intervening path. From equation (1.6)

$$dG = dE + P\,dV + V\,dP - T\,dS - S\,dT \tag{1.7}$$

and when T and P are constant

$$(dG)_{T,P} = dE + P\,dV - T\,dS \tag{1.8}$$

We now (arbitrarily) divide the total work done by (or on) a system into "pressure × volume work" and "useful work" so that

$$dW = P\,dV + dW' \tag{1.9}$$

where dW' represents the useful work and, as before, is positive when the system performs work on its surroundings and is negative when work is performed on the system.[6]

Substituting equation (1.9) into equation (1.1), we obtain

$$dE = dQ - P\,dV - dW' \tag{1.10}$$

and substituting equation (1.10) into equation (1.8) we obtain

$$(dG)_{T,P} = dQ - T\,dS - dW' \tag{1.11}$$

Since $dS = d_e S + d_i S = (dQ/T) + d_i S$,

$$T\,dS = dQ + T\,d_i S$$

so that from equation (1.11) we obtain

$$-(dG)_{T,P} = T\,d_i S + dW' \tag{1.12}$$

Thus:

(a) If a system at constant T and P performs *useful work* on its surroundings by means of a reversible process, $d_i S = 0$ and the decrease in the Gibbs free energy of the system ($-dG$) is equal to the useful work performed. If the process is irreversible, $d_i S > 0$ and the decrease in G is *greater* than the useful work performed.

(b) If a spontaneous process occurs in a closed system (one that cannot exchange matter with its environment), $dW' = 0$ and the decrease in free energy of the system is reflected only by an increase in internal entropy; that is, free energy is completely dissipated.

These considerations provide some insight into the meanings of internal energy, free energy, and entropy. Consider for the moment an irreversible process that takes place in a closed system (e.g., the diffusion of a solute from a region of higher concentration to one of lower concentration). Clearly, the internal energy of the system will not be affected since dQ and dW are both zero. However, the free energy of the system will decrease and reach zero when equilibrium is achieved. Thus, the potential ability of the system to perform work is dissipated at the expense of the production of entropy; for this reason free energy functions are frequently referred to as "work functions." The same holds for a system that can perform work on its surroundings. For any real process the decrease in free energy is greater than the amount of useful or recoverable work performed. Thus, the free energy of the universe has decreased but the internal energy is conserved.

Why? At the outset it should be stressed that thermodynamics is not based on any particular model of matter, and the functions E and S were introduced as properties of state that simply permit a consistent description of the energetics and direction of natural events. We cannot search for a more profound interpretation of these functions in classical thermodynamics. However, let us depart from this secure axiomatic approach and consider a closed system subdivided into two compartments by a porous barrier with one compartment containing a monoatomic gas and the other initially void. Clearly, the gas will diffuse (expand) into the initially empty compartment and equilibrium will be achieved when the distribution of atoms is random (i.e., equal concentrations in both compartments). Now, the internal energy of this system is simply the sum of the kinetic, translational energies of the individual atoms and is given by $E = n \cdot m \cdot \bar{v}^2/2$ where n is the number of atoms, m is the atomic mass and $\bar{v}^2$ is the mean square velocity of each atom that is only dependent on T. It should be noted that although $\bar{v}$ is a vector, $\bar{v}^2$ is a scalar, so that while E embodies the concept of motion, it is divorced from the notion of direction. E would be the same if at any instant every atom were moving in the same direction as when the movements are directed randomly. In contrast, the entropy function is a measure of the randomness of the system. An increase in entropy or an increase in randomness may be defined as an increase in the number of possible configurations or complexions of a system, all of which are equally likely and accessible; the state of equilibrium represents the most random or most probable configuration that the system can achieve. Thus, the second law states that the direction of all spontaneous processes is from a less random (less probable) to a more random (more probable) state. Clearly, with an increase in the randomness of a system, the ability of the system to convert its internal energy to a directed effort, or useful work, decreases.[7]

At the risk of belaboring this point, let us close this discussion with a trivial example. When a heavy weight is dropped onto a metal plate in a vacuum, the kinetic energy of the falling weight is completely converted into thermal energy, which in turn is reflected by an increase in the velocity of random motion of particles (entropy). If the motion were not random, but instead if all of the particles could be harnessed to simultaneously move upward against the weight, the weight would rise and a spontaneous process will have been reversed without outside intervention. The second law infers that the *probability* of this occurrence is so low that this event may be considered impossible.[8]

1.4 The electrochemical potential

We now employ the Gibbs free energy function to derive an expression for the electrochemical potential of a substance i. As we will see, the electrochemical potential is an intensive property of a system, and a difference in electrochemical potential is *the* conjugate driving force for the diffusional flow of matter. This property will play a central role in all of our subsequent considerations so that the importance of understanding its meaning cannot be overstated.

Let us start by considering the effect of reversibly transferring matter into a system on G. Inasmuch as

$$G = E + PV - TS \tag{1.6}$$

$$dE = dQ - P\,dV - dW' \tag{1.10}$$

and for a reversible process

$$dQ = T\,dS$$

it is a simple matter to show that

$$dG = -S\,dT + V\,dP - dW' \tag{1.13}$$

where dW' is the reversible work of transfer of matter into the system given by $\Sigma_i \mu_i \, dn_i + \psi \, de$ [equation (1.5)].

Since G is a property of state and, as such, is a function of other state properties such as T, P, and composition, we can express dG as an exact differential of these other properties of state as follows:

$$dG = \left(\frac{\partial G}{\partial T}\right)_{P,n_i,n_j,\ldots} dT + \left(\frac{\partial G}{\partial P}\right)_{T,n_i,n_j,\ldots} dP + \sum_i \left(\frac{\partial G}{\partial n_i}\right)_{T,P,n_j,\ldots} dn_i \tag{1.14}$$

where $n_i, n_j, \ldots$ represent the number of moles of $i, j, \ldots$ in the system. Comparison of equation (1.14) with equation (1.13) discloses that

$$\left(\frac{\partial G}{\partial T}\right)_{P,n_i,n_j,\ldots} = -S; \quad \left(\frac{\partial G}{\partial P}\right)_{T,n_i,n_j,\ldots} = V \tag{1.15}$$

and

$$-dW' = \sum_i \left(\frac{\partial G}{\partial n_i}\right)_{T,P,n_j,\ldots} dn_i \tag{1.16}$$

Since

$$dW' = \sum_i \mu_i \, dn_i + \psi \, de$$

and

$$de = z_i \mathcal{F} \, dn_i$$

where z_i is the valence of i and $\mathcal{F}$ is the Faraday, it follows that

$$dW' = \sum_i (\mu_i + z_i \mathcal{F} \psi) \, dn_i$$

We now define the electrochemical potential of i as

$$\tilde{\mu}_i = \mu_i + z_i \mathcal{F} \psi \tag{1.17}$$

so that

$$dW' = \sum_i \tilde{\mu}_i \, dn_i \tag{1.18}$$

Thus, from equation (1.16) it is clear that $\tilde{\mu}_i = (\partial G_i / \partial n_i)$ or the *partial molal free energy* of i and represents the increase in free energy of the system resulting from the addition of one mole of i when T, P, and the composition of all other substances $n_j \ldots$ are constant. In other words, it is the work that must be performed on the system in order to introduce 1 mole of i when all other properties of state are maintained constant. Clearly, $\Delta\tilde{\mu}_i$ is the conjugate driving force for the flow of i.

It should be noted that the electrochemical potential $\tilde{\mu}_i$ combines the partial molal free energy change resulting from a change in the chemical composition of the system (the chemical potential μ_i) and (if $z_i \neq 0$) from the addition of charge to a system whose electrical potential is ψ.

We now wish to define $\tilde{\mu}_i$ explicitly in terms of measurable parameters. Clearly, $\tilde{\mu}_i$ is an intensive property of state since it is the ratio of two extensive properties [equation (1.16)] so that $d\tilde{\mu}_i$ can be expressed as an exact differential of all other intensive properties such as temperature (T), pressure (P), concentrations ($c_i, c_j, \ldots$), and electrical potential (ψ). Thus,

$$d\tilde{\mu}_i = \left(\frac{\partial \tilde{\mu}_i}{\partial T}\right)_{P,\psi,c_i,c_j,\ldots} dT + \left(\frac{\partial \tilde{\mu}_i}{\partial P}\right)_{T,\psi,c_i,c_j,\ldots} dP + \left(\frac{\partial \tilde{\mu}_i}{\partial \psi}\right)_{T,P,c_i,c_j,\ldots} d\psi + \left(\frac{\partial \tilde{\mu}_i}{\partial c_i}\right)_{T,P,\psi,c_j,\ldots} dc_i + \sum_{j \neq i} \left(\frac{\partial \tilde{\mu}_i}{\partial c_j}\right)_{T,P,\psi,c_i,\ldots} dc_j + \cdots \tag{1.19}$$

Now, from equations (1.15) and (1.16)

$$\left(\frac{\partial \tilde{\mu}_i}{\partial T}\right) = \frac{\partial}{\partial T}\left(\frac{\partial G}{\partial n_i}\right) = \frac{\partial}{\partial n_i}\left(\frac{\partial G}{\partial T}\right) = -\left(\frac{\partial S}{\partial n_i}\right) = -\bar{s}_i \tag{1.20}$$

and

$$\left(\frac{\partial \tilde{\mu}_i}{\partial P}\right) = \frac{\partial}{\partial P}\left(\frac{\partial G}{\partial n_i}\right) = \frac{\partial}{\partial n_i}\left(\frac{\partial G}{\partial P}\right) = \frac{\partial V}{\partial n_i} = \bar{v}_i \tag{1.21}$$

where $\bar{s}_i$ is the partial molal entropy of i and $\bar{v}_i$ is the partial molal volume of i.

Further, from equation (1.17)

$$\left(\frac{\partial \tilde{\mu}_i}{\partial \psi}\right)_{T,P,c_i,c_j,\ldots} = z_i \mathcal{F} \tag{1.22}$$

Now we are close to an explicit statement of equation (1.19), and it only remains to define the relations between $\tilde{\mu}_i$ and $c_i, c_j, \ldots$.

The relation between $\tilde{\mu}_i$ and c_i stems from the brilliant insights made by some of the pioneers of physical chemistry around the turn of the nineteenth century-Henry, Raoult, DeVries, Pfeffer, Hamburger, Arrhenius, and van't Hoff-which led to the notion that the behavior of solutes in dilute solutions resembles the behavior of a perfect gas! Thus, in 1886 van't Hoff wrote: "The pressure which a gas exerts when a given number of molecules are distributed in a given volume is equally great as the osmotic pressure, which under the same conditions would be produced by most solutes when they are dissolved in an arbitrary solvent."

Thus, in analogy with the perfect gas law, we can write

$$P_i = n_i RT/V = c_i RT \tag{1.23}$$

where P_i may be considered the "partial pressure," "fugacity," or "escaping tendency" (cf. Lewis and Randall, 1961) of a solute in dilute solution.

Now,

$$\left(\frac{\partial \tilde{\mu}_i}{\partial c_i}\right) = \left(\frac{\partial \tilde{\mu}_i}{\partial P_i}\right) \cdot \left(\frac{\partial P_i}{\partial c_i}\right)$$

and, from equation (1.23)

$$\left(\frac{\partial P_i}{\partial c_i}\right) = RT$$

so that

$$\left(\frac{\partial \tilde{\mu}_i}{\partial c_i}\right) = RT \cdot \left(\frac{\partial \tilde{\mu}_i}{\partial P_i}\right)$$

Also,

$$\left(\frac{\partial \tilde{\mu}_i}{\partial P_i}\right) = \left(\frac{\partial \tilde{\mu}_i}{\partial P}\right) \cdot \left(\frac{\partial P}{\partial P_i}\right)$$

From equation (1.21), $(\partial\tilde{\mu}_i/\partial P) = \overline{v}_i$ and from Henry's law $P = P_i + \Sigma_{j \neq i} P_j$, so that $(\partial P/\partial P_i) = 1$.

Therefore, $(\partial\tilde{\mu}_i/\partial P_i) = \overline{v}_i$; and since $\overline{v}_i = (\partial V/\partial n_i) = (RT/P_i) = 1/c_i$, we may finally write

$$\left(\frac{\partial \tilde{\mu}_i}{\partial c_i}\right)_{T,P,\psi,c_j,\ldots} = \frac{RT}{c_i} = \frac{RT\, d\ln c_i}{dc_i} \tag{1.24}$$

Now, what about the term $\Sigma_{j \neq i}(\partial\tilde{\mu}_i/\partial c_j)\, dc_i$? For an ideal, infinitely dilute system (where the perfect gas law holds) the electrochemical potential of i is not affected by the presence of other particles (j) so that this term can be neglected; thus, we can combine the results of equations (1.19) to (1.24) and obtain

$$d\tilde{\mu}_i = -\overline{s}_i\, dT + \overline{v}_i\, dP + z_i \mathcal{F}\, d\psi + RT\, d\ln c_i \tag{1.25}$$

An explicit expression for $\tilde{\mu}_i$ can be obtained by integrating equation (1.25) from $T = 0$ to $T = T$, $P = 0$ to $P = P$, $\psi = 0$ to $\psi = \psi$, and $c_i = 1$ to $c_i = c_i$, that is,

$$\tilde{\mu}_i = \mu_i^0 - \overline{s}_i T + \overline{v}_i P + z_i \mathcal{F}\psi + RT \ln c_i \tag{1.26}$$

where the constant of integration, μ_i^0, is the *standard electrochemical potential* and is a function of the solvent alone.[9]

Under isothermal conditions, $dT = 0$, so that integration of equation (1.25) yields

$$\tilde{\mu}_i = (\mu_i{}^0)_T + \bar{v}_i P + z_i \mathcal{F} \psi + RT \ln c_i \tag{1.27}$$

where $(\mu_i{}^0)_T$ is a function of the nature of the solvent and T.

Under isothermal and isobaric conditions, $dT = dP = 0$, so that

$$\tilde{\mu}_i = (\mu_i{}^0)_{T,P} + z_i \mathcal{F} \psi + RT \ln c_i \tag{1.28}$$

where $(\mu_i{}^0)_{T,P}$ is a function of the solvent, T and P.

The practical significance of equations (1.25) to (1.28) is that they provide the means for calculating the difference in $\tilde{\mu}_i$ across a membrane separating two homogeneous compartments, designated by the superscripts o and i, which have a common solvent. Thus, in general,

$$\Delta\tilde{\mu}_i = (\tilde{\mu}_i)^o - (\tilde{\mu}_i)^i = -\bar{s}_i(T^o - T^i) + \bar{v}_i(P^o - P^i) + z_i\mathcal{F}(\psi^o - \psi^i) + RT \ln \left(\frac{c_i{}^o}{c_i{}^i}\right) \tag{1.29}$$

Simplifications of equation (1.29) for specific conditions, for example, $T^o = T^i$, $P^o = P^i$, are self-evident.

In developing these expressions for $\tilde{\mu}_i$, we assumed that the composition of the system is sufficiently dilute so that the individual particles (molecules, atoms, ions) of i do not interact with each other or with those of other components $j, k, l, \ldots$. Two problems arise if this *ideal* condition does not pertain.

First, the perfect gas law analogy $c_i = (P_i/RT)$ cannot be applied. However, instead we can write

$$a_i = \gamma_i c_i = P_i/RT$$

where a_i is the *activity* of i and γ_i is the *activity coefficient.* Then,

$$\mu_i = \mu_i{}^0 + RT \ln a_i = \mu_i{}^0 + RT \ln c_i + RT \ln \gamma_i$$

and

$$\left(\frac{\partial \mu_i}{\partial c_i}\right) = \frac{RT}{c_i} + RT \left(\frac{\partial \ln \gamma_i}{\partial c_i}\right)$$

The second problem arises from the fact that in a nonideal multicomponent solution, μ_i is influenced by $c_j, c_k, \ldots$ so that $\Sigma\,(\partial\mu_i/\partial c_j) \neq 0$. These effects can be included in the activity coefficient γ_i. Thus, in general, we can write

$$d\mu_i = RT\, d \ln a_i = RT\, d \ln c_i + RT\, d \ln \gamma_i$$

where

$$d \ln \gamma_i = \sum_{i}^{n} \left(\frac{\partial \ln \gamma_i}{\partial c_i} \right) dc_i \tag{1.30}$$

Equation (1.30) includes all of the effects of c_i, c_j, c_k, and so forth on γ_i.

Let us consider the spontaneous (irreversible) diffusion of dn_i moles of i from compartment o to compartment i. Clearly, $dG^o = -(\tilde{\mu}_i)^o\, dn_i$ and $dG^i = (\tilde{\mu}_i)^i\, dn_i$. Since G is an extensive property of state, $dG = dG^o + dG^i = [(\tilde{\mu}_i)^i - (\tilde{\mu}_i)^o]\, dn_i$. According to equation (1.12), when an irreversible process takes place in a closed system, $dW' = 0$ and $dG < 0$. It follows that in order for this transfer to occur spontaneously, $(\tilde{\mu}_i)^o$ must be greater than $(\tilde{\mu}_i)^i$. *Thus, in a system characterized by a single force* ($\Delta\tilde{\mu}_i$) *and a single flow* (dn_i), *the flow must be directed from a region of higher electrochemical potential to one of lower electrochemical potential.* Further, since the compartments must be finite, as the flow continues, $\Delta\tilde{\mu}_i \to 0$ and $dG \to 0$. When $\Delta\tilde{\mu}_i = 0$, then $dG = 0$ and the system will have reached equilibrium.

Finally, according to equation (1.12), when an irreversible process takes place in a closed system at constant T and P

$$-dG = T\, d_iS$$

From these considerations,

$$-dG = [(\tilde{\mu}_i)^o - (\tilde{\mu}_i)^i]\, dn_i = \Delta\tilde{\mu}_i\, dn_i$$

Thus,

$$T\, d_iS = \Delta\tilde{\mu}_i \cdot dn_i$$

Dividing by dt, we obtain $T\, d_iS/dt = J_i \cdot \Delta\tilde{\mu}_i$, where $J_i = dn_i/dt$ and is the flux of i in moles per unit time.

Thus, when an irreversible process takes place in a closed system at constant T and P, the Gibbs free energy is dissipated at the expense of internal entropy production. The rate of formation of internal entropy (or dissipation of free energy) is proportional to the product of the flow (the rate of displacement of an extensive property) and the conjugate driving force (a difference in the conjugate intensive property).

1.5. Conditions of equilibrium

As discussed previously, when a system is undergoing a reversible change or passing through a series of equilibrium states, $dS = dQ/T$ so that $d_iS = 0$. From equation (1.12) we see that in a closed system at constant temperature and pressure $-dG = T\, d_iS/dt$. It follows that in such a system the criterion for equilibrium is $dG = 0$. If the system consists of two well-stirred solutions containing n solutes separated by an inert membrane at constant T and P, it will be at equilibrium when all diffusional flows cease, that is, when $\Delta\tilde{\mu}_i = 0$ for all species i that can permeate the membrane. Thus, from equation (1.28) we see that any species i will be at equilibrium when

$$\Delta\tilde{\mu}_i = (\Delta\mu_i^0)_{T,P} + z_i \mathcal{F}\Delta\psi + RT\Delta \ln c_i = 0$$

Let us distinguish between the two solutions using the superscripts o and i. It follows that for an uncharged solute, $\Delta\tilde{\mu}_i = 0$ when

$$\left(\frac{c_i^o}{c_i^i}\right) = \exp\left[\frac{(\mu_i^0)^i - (\mu_i^0)^o}{RT}\right] \tag{1.31}$$

where, as noted, the standard state chemical potential μ_i^0 is a function of the nature of the solvent, T and P. Thus, if both solutions have the same solvent, $\Delta\tilde{\mu}_i = 0$ when $(c_i^o/c_i^i) = 1$. On the other hand, if the solvents differ and i is more soluble in one than in the other, at equilibrium $(c_i^o/c_i^i) \neq 1$. The value of (c_i^o/c_i^i) is frequently referred to as the "partition coefficient" (β_i); it will be considered further in Section 2.1 with respect to the distribution of an uncharged solute across the interfaces between a membrane and the adjacent solutions. Clearly, if i is more soluble (more stable) in the solvent in compartment i than in the solvent in compartment o, $(\mu_i^0)^i < (\mu_i^0)^o$ so that, at equilibrium, $c_i^i > c_i^o$.

If i is charged, $\Delta\tilde{\mu}_i = 0$ when

$$\left(\frac{c_i^o}{c_i^i}\right) = \exp\left\{\left[\frac{(\mu_i^0)^i - (\mu_i^0)^o}{RT}\right] + \left[\frac{z_i\mathcal{F}(\psi^i - \psi^o)}{RT}\right]\right\}$$

Thus, if the two solutions have the same solvent properties with respect to i, $[(\mu_i^0)^i - (\mu_i^0)^o] = 0$ and

$$\left(\frac{c_i^o}{c_i^i}\right) = \exp\left[\frac{z_i\mathcal{F}(\psi^i - \psi^o)}{RT}\right]$$

or

$$\Delta\psi = (\psi^i - \psi^o) = \left(\frac{RT}{z_i\mathcal{F}}\right) \ln \left(\frac{c_i^o}{c_i^i}\right) \tag{1.32}$$

This relation is referred to as the *Nernst equilibrium equation* and $\Delta\psi$ is frequently referred to as the *Nernst potential*. It will be discussed further in a variety of contexts, but in essence $\Delta\psi$ is the electrical potential difference necessary to balance (i.e., equivalent to) the concentration ratio of a charged species so that the total driving force $(\Delta\tilde{\mu}_i)$ is zero.

1.6. Irreversible (nonequilibrium) thermodynamics

Classical thermodynamics derives its inviolable authority from the fact that it is divorced from any considerations of the structure of matter and it confines its attention to systems at equilibrium or undergoing idealized, reversible changes (which are, in fact, an infinite sequence of equilibrium states). Under these conditions the properties of state are well defined and uniform throughout the system. All *real* processes are described by a set of inequalities that simply indicate the direction of change, but there is no concern with the rate of change and, indeed, the discipline does not formally recognize the element of *time.* Although the amount of useful information that has resulted (and will continue to emerge) from the axiomatic application of two postulates boggles the mind, it is clear that, because of its limited scope, classical thermodynamics is not ideally suited for the description of living systems that are displaced from equilibrium and that are characterized by flows of matter and energy within the system as well as between the system and its surroundings.

Irreversible or nonequilibrium thermodynamics is a relatively recent outgrowth of efforts to extend the principles of classical thermodynamics to systems displaced from equilibrium that are characterized by irreversible flows of matter and energy.[10] It succeeds in replacing the inequalities of classical thermodynamics with equalities and, by explicitly introducing the element of time, it deals with rates or flows. However, there is a price to be paid inasmuch as the principles of irreversible thermodynamics are not generally valid and the range of validity must be established (where possible) empirically; this new discipline does not command the unquestionable authority of its forerunner.

The linear phenomenological equations

For many years it has been recognized on empirical grounds that in a system characterized by a single force and a single flow, the flow is a linear function of the force over a relatively large range of flows and forces. Thus, we can write

$$J_i = L_{ii} X_i \tag{1.33}$$

where J_i is the flow of an extensive property, X_i is the conjugate driving force (a difference in the conjugate intensive property), and L_{ii} is a proportionality constant having units of conductance. Ohm's law of current flow, Fourier's law of heat flow, Fick's law of diffusion, and the Poiseuille's equation describing volume flow are familiar examples of equation (1.33).

It is also known that if a system is characterized by several flows and forces, there *may* be interactions (coupling) between nonconjugate flows and forces.

Further, for sufficiently slow flows in a system that is not displaced *too* far from equilibrium, the dependence of flows on nonconjugate forces is also linear. These empirical observations can be described by a set of linear phenomenological equations as follows:

$$\begin{aligned}
J_i &= L_{ii}X_i + L_{ij}X_j + L_{ik}X_k + \cdots + L_{in}X_n \\
J_j &= L_{ji}X_i + L_{jj}X_j + L_{jk}X_k + \cdots + L_{jn}X_n \\
J_k &= L_{ki}X_i + L_{kj}X_j + L_{kk}X_k + \cdots + L_{kn}X_n \\
&\vdots \\
J_n &= L_{ni}X_i + L_{nj}X_j + L_{nk}X_k + \cdots + L_{nn}X_n
\end{aligned} \tag{1.34}$$

or

$$J_i = L_{ii}X_i + \sum_j^n L_{ij}X_j$$

In these equations the coefficients L_{ii}, L_{jj}, L_{kk}, and so on are referred to as *straight coefficients* inasmuch as they relate the flow with its conjugate driving force. The $L_{ij(j \neq i)}$ are referred to as *cross coefficients* or *coupling coefficients* inasmuch as they relate flows to nonconjugate driving forces.

It is also possible to express the forces in terms of linear combinations of flows, as follows:

$$\begin{aligned}
X_i &= R_{ii}J_i + R_{ij}J_j + R_{ik}J_k + \cdots + R_{in}J_n \\
X_j &= R_{ji}J_i + R_{jj}J_j + R_{jk}J_k + \cdots + R_{jn}J_n \\
X_k &= R_{ki}J_i + R_{kj}J_j + R_{kk}J_k + \cdots + R_{kn}J_n \\
&\vdots \\
X_n &= R_{ni}J_i + R_{nj}J_j + R_{nk}J_k + \cdots + R_{nn}J_n
\end{aligned} \tag{1.35}$$

or

$$X_i = R_{ii}J_i + \sum_j^n R_{ij}J_j$$

where the R's are now generalized resistances.

It should be noted that the L's and R's are, in general, functions of the state of the system but are independent of the forces. We will encounter specific situations later where the L's or R's are functions of concentration but are independent of concentration or electrochemical potential *differences.*

The choice of equations (1.34) *or* (1.35) for the description of a given system is a matter of preference or convenience. Some investigators prefer the use of resistances since (as will be shown) these can be given physical expression in terms of frictional interactions. It should be noted, however, that we can always convert one system of coefficients into the other using the relation

$$R_{ik} = \frac{|L_{ik}|}{|L|}$$

where $|L|$ is the determinant of the matrix of the coefficients L_{ik} and $|L_{ik}|$ is the minor of the matrix corresponding to the term L_{ik}. Thus, for a system with only two flows (J_i and J_j) and two forces (X_i and X_j)

$$R_{ii} = \left(\frac{L_{jj}}{|L|}\right); \quad R_{ij} = \left(\frac{-L_{ij}}{|L|}\right); \quad R_{ji} = \left(\frac{-L_{ji}}{|L|}\right)$$

and

$$R_{jj} = \left(\frac{L_{ii}}{|L|}\right)$$

The "dissipation" function

A fundamental assumption that underlies the entire framework of irreversible thermodynamics is that the total change in the entropy of a system is given by the Gibbs equation (1.5) even though the system is not at equilibrium and is undergoing an irreversible change. (Recall, that the substitution of $T\,dS$ for dQ in deriving the Gibbs equation is strictly valid only for a reversible process.) This assumption has a limited range of validity and can only be employed when the system "is not too far" from equilibrium. In the final analysis, the justification for this approach in a given instance can only be established by the experimental validation of the theoretical predictions.

If the system is not displaced *too* far from equilibrium so that the total change in entropy is given by the Gibbs equation, then using the relation $dS = d_eS + d_iS$, it can be shown that

$$T\left(\frac{d_iS}{dt}\right) = \sum_i^n J_iX_i \tag{1.36}$$

where J_i and X_i are the appropriate thermodynamic conjugates (i.e., Table 1.1). Thus, the inequality ($d_iS > 0$) of classical thermodynamics is replaced by a definable equality. The expression $T(d_iS/dt)$ is sometimes symbolized by Φ and is referred to as the *dissipation function.* As discussed before, in a closed system at constant T and P, this function is equal to the rate of decrease (dissipation) of the Gibbs free energy ($-dG/dt$); more generally, it can always be related to the rate of dissipation of free energy or the rate of decrease in the ability to perform useful work.

An intuitive appreciation of the dissipation function can be obtained by considering a simple system in which there is only a single flow and a single force. Then, $X_i = R_{ii}J_i$ so that

$$T\left(\frac{d_iS}{dt}\right) = R_{ii}J_i^2$$

Clearly, since $T(d_iS/dT)$ is always positive, R_{ii} must also be positive and *independent of the direction of the flow.* Obviously, if in our example J_i is the flow of electrical current I across a resistor R driven by a potential difference,

$$T\left(\frac{d_iS}{dt}\right) = I^2R$$

which is immediately recognized as the rate at which power is dissipated (heat is generated) by such a system.

An important corollary follows; namely, in a system in which there is only a single force and a single flow, all of the free (potential) energy is dissipated. Useful work can only be accomplished by coupling one flow to another.

Onsager's reciprocal relations

In two fundamental papers published in 1931, Onsager interwove the linear phenomenologic equations (which were generated by empirical observations) and the dissipation function (whose derivation involves a questionable assumption) in a brilliant synthetic move that immeasurably strengthened the theoretical credibility of the irreversible thermodynamic approach; this achievement and others were recognized by the award of the Nobel Prize in chemistry in 1968.

Onsager demonstrated that if the flows and forces of equations (1.34) are chosen appropriately (i.e., the correct thermodynamic conjugates) so that equation (1.36) is satisfied then

$$L_{ij} = L_{ji} \qquad \text{for all } i \text{ and } j \tag{1.37}$$

One important consequence of these "reciprocal relations" is that they permit a significant reduction in the number of coefficients that must be experimentally determined. The number of coefficients needed to completely describe a system with n flows and n forces is n^2, so that the evaluation of these coefficients becomes rather formidable even for relatively simple systems (i.e., $n = 3$). When Onsager's relations apply, the number of coefficients is reduced to $n(n + 1)/2$. Another important consequence is that the experimental demonstration of reciprocity for a given system supports the contention that the system is not too far displaced from equilibrium, so that it can be described by a combination of linear relations between flows and forces, where the sum of the products of conjugate flows and forces is a measure of the rate of internal entropy production. Equation (1.37) has been experimentally verified for a number of systems, including diffusion in ternary systems (e.g., NaCl—KCl—H_2O) and electrokinetic phenomena (see Section 4.4). Some of these systems were characterized by rather large forces, suggesting that this approach has a wide range of validity and that the admonition that the system cannot be too far displaced from equilibrium may not be overly restrictive (cf. Dunlop, 1957; Dunlop and Gosting, 1959; Fujita and Gosting, 1960; Miller, 1960).

Finally, an additional constraint that emerges from these considerations stems from the *positive definite* nature of $T(d_iS/dt)$. Since the dissipation function can never be negative, the straight coefficients must be positive and the cross coefficients must satisfy the condition

$$L_{ij}^2 > L_{ii}L_{jj} \tag{1.38}$$

Coupled flows, energy conversion, and useful work

Finally, let us briefly consider a subject that will be treated in greater detail in Chapters 5 and 6. Consider a system in which there are two forces and two flows that can be described as follows:

$$\begin{aligned} J_i &= L_{ii}X_i + L_{ij}X_j \\ J_j &= L_{ij}X_i + L_{jj}X_j \end{aligned} \tag{1.39}$$

If $L_{ij} = 0$, the flows are uncoupled and

$$T\left(\frac{d_iS}{dt}\right) = J_iX_i + J_jX_j = L_{ii}X_i^2 + L_{jj}X_j^2 = R_{ii}J_i^2 + R_{jj}J_j^2 > 0$$

In this system, all of the free energy stored in the forces X_i and X_j will eventually be dissipated.

However, if $L_{ij} < 0$ and if $|L_{ij}X_j| > L_{ii}X_i$, then $J_i < 0$. This is an example of *negative coupling*, where the nonconjugate force X_j can drive J_i in a direction *opposite* to the direction of its conjugate driving force and thereby perform useful work. The dissipation function under these conditions is given by

$$T\left(\frac{d_iS}{dt}\right)' = J_jX_j + J_iX_i = L_{ii}X_i^2 + L_{jj}X_j^2 + 2L_{ij}X_iX_j > 0$$

Obviously, since $J_i < 0$ and $L_{ij} < 0$, $T(d_iS/dt)'$ is less than $T(d_iS/dt)$ and the difference is the rate at which free energy is converted into useful work rather than being simply dissipated.

The example cited is one in which energy from a flow in the direction of its conjugate driving force is converted into useful work by a (negative) coupling mechanism that results in the propulsion of a second flow in the direction opposite to its conjugate driving force (e.g., the flow of matter against a difference in electrochemical potential or volume against a difference in pressure). The efficiency of energy conversion is given by

$$\eta = \frac{\text{power output}}{\text{power input}} = \left(\frac{-J_i X_i}{J_j X_j}\right) = 1 - T\left(\frac{d_i S}{dt}\right) \tag{1.40}$$

For an ideal reversible process, $d_i S/dt = 0$ and $\eta = 1$.

2 Isothermal diffusion

This chapter deals with the diffusion of a solute across a barrier separating two homogeneous solutions in the same solvent and at the same temperature. The notation that will be employed is illustrated in Figure 2.1. The superscripts o and i refer to the two surrounding solutions and the overbars designate intramembrane properties. The bidirectional fluxes of a solute i are designated J_i^{oi} and J_i^{io}, and the net flux is given by $J_i = J_i^{oi} - J_i^{io}$.[1] For purposes of simplification we will refer to concentrations (c_i) and implicitly assume that $\gamma_i = 1$ (or that $\gamma_i^o = \gamma_i^i$); however, it is important to stress that, strictly speaking, activities should be employed rather than concentrations. Finally, we will assume that each solution is perfectly stirred and ignore problems that can arise because of unstirred layers.[2]

2.1. The Nernst-Planck equation

Consider the system illustrated in Figure 2.2, where 1 cm^3 of solution containing n solute particles lies adjacent to 1 cm^2 of membrane surface. Clearly, the number of particles that will cross the membrane in the x direction per unit time is simply

$$J_i = c_i v_i$$

where v_i is the velocity of each solute particle. If v_i is expressed in centimeters per second and c_i in moles per cubic centimeter, then J_i has units of moles per centimeter squared, second; when $v_i = 1$ cm/sec all of the particles will have crossed from left to right in 1 sec. Now, v_i is directly proportional to the force acting on each particle (f_i) and is given by $v_i = f_i u_i$ where the proportionality constant u_i is defined as the mobility and is simply the velocity per unit force.[3] Thus, we can write the general expression

$$J_i = c_i u_i f_i \tag{2.1}$$

In general, *force* may be defined as the rate of change of *energy* with *distance* (dE/dx) so that the force acting on a mole of matter is simply the gradient of the free energy per mole or the gradient of the electrochemical poten-

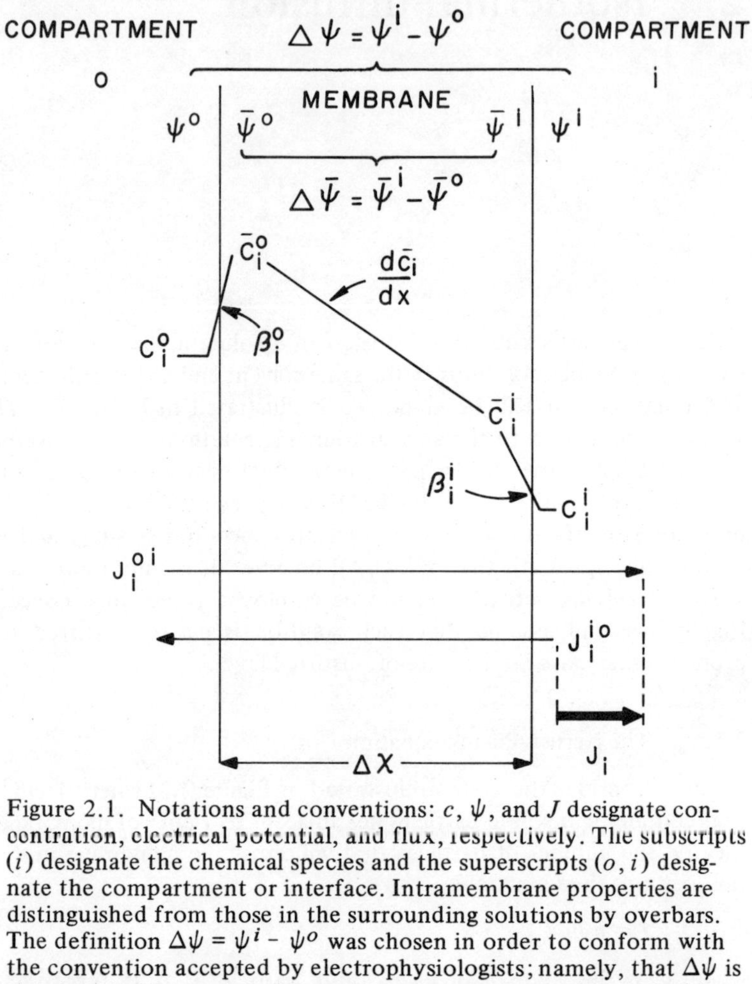

Figure 2.1. Notations and conventions: c, ψ, and J designate concentration, electrical potential, and flux, respectively. The subscripts (i) designate the chemical species and the superscripts (o, i) designate the compartment or interface. Intramembrane properties are distinguished from those in the surrounding solutions by overbars. The definition $\Delta\psi = \psi^i - \psi^o$ was chosen in order to conform with the convention accepted by electrophysiologists; namely, that $\Delta\psi$ is the electrical potential of the cell interior or inner solution bathing epithelial membranes with respect to the extracellular or outer solution.

tial. Thus, we conclude that

$$J_i = -c_i u_i (d\tilde{\mu}_i / dx) \tag{2.2}$$

Equation (2.2) is referred to as the Nernst–Planck equation and is the most common starting point for the formal description of diffusion. The reason for the negative sign is that J_i is defined as positive when the flow takes place from a region of high electrochemical potential to one of lower electrochemical potential, so that a *positive* flow is driven by a *negative* electrochemical

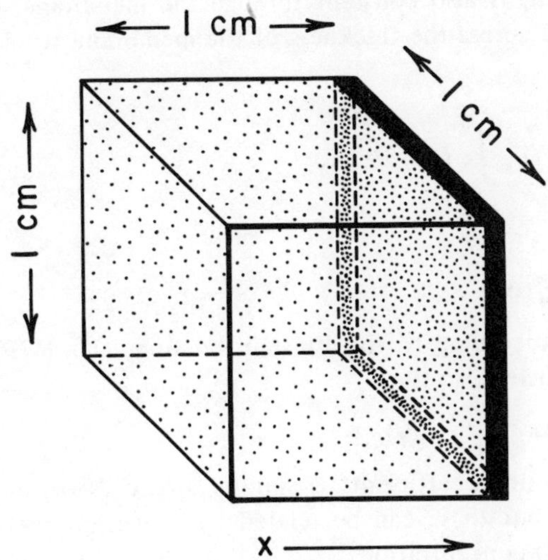

Figure 2.2. A 1 cm^3 volume of solution adjacent to a membrane with a surface area of 1 cm^2.

potential gradient. It should be emphasized at this point that equation (2.2) assumes that J_i is driven entirely by its conjugate driving force and is not affected by other flows or forces. In particular, equation (2.2) assumes that there is no convective flow of i resulting from interaction with the flow of solvent; this matter is discussed in detail by Helfferich (1962) and will be considered in Section 4.4.

For a system at constant temperature

$$\tilde{\mu}_i = (\mu_i{}^0)_T + \bar{v}_i P + RT \ln c_i + z_i \mathcal{F} \psi \qquad (2.3)$$

However, because $\bar{v}_i$ is small for most solutes, the term $\bar{v}_i P$ makes a very small contribution to $\tilde{\mu}_i$ and can often be neglected, particularly in biological systems,[4] so that

$$J_i = -u_i \bar{c}_i \,[RT(d \ln \bar{c}_i/dx) + z_i \mathcal{F}(d\bar{\psi}/dx)] \qquad (2.4)$$

Now, let us apply equation (2.4) to the description of the diffusion of an unchanged solute across a membrane whose thickness is Δx. Clearly, when $z_i = 0$, equation (2.4) is simply

$$J_i = -u_i \bar{c}_i RT(d \ln \bar{c}_i/dx) = -RTu_i(d\bar{c}_i/dx) \qquad (2.5)$$

where J_i is the flux of i in the x direction *at some point in the membrane*, u_i is the mobility of i *at that point*, $\bar{c}_i$ is the concentration of i *at that point*, and $(d\bar{c}_i/dx)$ is the concentration gradient *at that point*. When the system is in a steady state, J_i is constant and must be the same at all points within the membrane. Assuming that u_i is also constant through the membrane, equation (2.5) can be integrated across the thickness of the membrane from $x = o$ to $x = i$, that is,

$$J_i \int_o^i dx = -RTu_i \int_o^i (d\bar{c}_i/dx)\, dx$$

yielding

$$J_i = -RTu_i \,\Delta\bar{c}_i/\Delta x \qquad (2.6)$$

where $\Delta\bar{c}_i = \bar{c}_i^{\,i} - \bar{c}_i^{\,o}$. According to the Einstein relation, $D_i = RTu_i$, where D_i is the diffusion coefficient, so that

$$J_i = -D_i\,\Delta\bar{c}_i/\Delta x = -P_i'\,\Delta\bar{c}_i \qquad (2.7)$$

where P_i', the permeability coefficient, is simply $D_i/\Delta x$. Now, in general, $\bar{c}_i^{\,o} \neq c_i^{\,o}$ and $\bar{c}_i^{\,i} \neq c_i^{\,i}$, but these can be related by "partition coefficients" β_i. If β_i is independent of concentration[5]

$$\beta_i = \bar{c}_i^{\,o}/c_i^{\,o} = \bar{c}_i^{\,i}/c_i^{\,i}$$

and equation (2.7) can be written in terms of the concentrations in the external solution as follows

$$J_i = -D_i\beta_i\,\Delta c_i/\Delta x = -P_i\,\Delta c_i \qquad (2.8)$$

where $\Delta c_i = c_i^{\,i} - c_i^{\,o}$ and P_i now includes the partition coefficient and is simply $\beta_i P_i'$. Equation (2.8) is the statement of Fick's *first law of diffusion* for a planar barrier under steady-state conditions.

2.2. The constant-field equation

When the solute is charged ($z_i \neq 0$) and there is an electrical potential difference across the membrane, the situation is far more complex. Under these conditions

$$J_i = -u_i\bar{c}_i\left[RT\left(\frac{d\ln\bar{c}_i}{dx}\right) + z_i\mathcal{F}\left(\frac{d\bar{\psi}}{dx}\right)\right]$$

or

$$J_i = -D_i\left[\frac{d\bar{c}_i}{dx} + \frac{\bar{c}_i z_i\mathcal{F}}{RT}\cdot\frac{d\bar{\psi}}{dx}\right] \qquad (2.9)$$

Multiplying both sides of equation (2.9) by $\exp(z_i\mathcal{F}\bar{\psi}/RT)$ and rearranging the right-hand side of the resulting expression yields

$$J_i\cdot\exp\left(\frac{z_i\mathcal{F}\bar{\psi}}{RT}\right) = -D_i\left\{\frac{d}{dx}\left[\bar{c}_i\exp\left(\frac{z_i\mathcal{F}\bar{\psi}}{RT}\right)\right]\right\} \qquad (2.10)$$

Assuming a steady state and that D_i is constant through the membrane, equation (2.10) can be integrated across the membrane thickness to give

$$J_i \int_o^i \exp\left(\frac{z_i \mathcal{F} \bar{\psi}}{RT}\right) dx = -D_i \int_o^i \left\{\frac{d}{dx}\left[\bar{c}_i \exp\left(\frac{z_i \mathcal{F} \bar{\psi}}{RT}\right)\right]\right\} dx$$

or

$$J_i \int_o^i \exp\left(\frac{z_i \mathcal{F} \bar{\psi}}{RT}\right) dx = -D_i \left[\bar{c}_i^{\,i} \exp\left(\frac{z_i \mathcal{F}\, \Delta\bar{\psi}}{RT}\right) - \bar{c}_i^{\,o}\right] \qquad (2.11)$$

where $\Delta\bar{\psi} = \bar{\psi}^i - \bar{\psi}^o$.

Obviously, this solution is incomplete because the expression on the left-hand side of equation (2.11) remains to be integrated. However, in order to accomplish this, the dependence of $\bar{\psi}$ on x within the membrane (i.e., the electrical potential profile) must be known or assumed. The simplest and most frequently used assumption is that suggested by Goldman (1943), namely, that $\bar{\psi}$ is a linear function of x or that $d\bar{\psi}/dx = \Delta\bar{\psi}/\Delta x$ ("the constant field assumption") so that $\bar{\psi}_x = \Delta\bar{\psi}(x/\Delta x)$, where x is some point in the membrane between 0 and Δx. The left-hand side of equation (2.11) then becomes

$$J_i \int_o^i \exp\left(\frac{z_i\, \mathcal{F} \Delta\bar{\psi}\, x}{RT\, \Delta x}\right) dx = J_i \left(\frac{RT\, \Delta x}{z_i \mathcal{F} \Delta\bar{\psi}}\right) \left[\exp\left(\frac{z_i \mathcal{F} \Delta\bar{\psi}}{RT}\right) - 1\right]$$

and the complete solution of equation (2.11) is

$$J_i = -\frac{D_i z_i \mathcal{F}\, \Delta\bar{\psi}}{RT\, \Delta x} \left[\frac{\bar{c}_i^{\,i} \exp(z_i\, \mathcal{F} \Delta\bar{\psi}/RT) - \bar{c}_i^{\,o}}{\exp(z_i\, \mathcal{F} \Delta\bar{\psi}/RT) - 1}\right] \qquad (2.12)$$

Equation (2.12) is frequently referred to as the Goldman equation or the *constant-field flux equation* and, as originally derived, referred only to intramembrane properties. In an effort to relate J_i to the measurable properties of the external solutions, Hodgkin and Katz (1949) assumed that the partition coefficient across the interfaces is concentration independent so that

$$\bar{c}_i^{\,o} = \beta_i c_i^{\,o} \quad \text{and} \quad \bar{c}_i^{\,i} = \beta_i c_i^{\,i}$$

Under these conditions, as will be discussed in Section 3.1, $(\bar{\psi}^o - \psi^o) = (\bar{\psi}^i - \psi^i)$ so that $(\bar{\psi}^i - \bar{\psi}^o) = (\psi^i - \psi^o) = \Delta\psi$, and

$$J_i = -\frac{P_i z_i\, \mathcal{F} \Delta\psi}{RT} \left[\frac{c_i^{\,o} - c_i^{\,i} \exp(z_i\, \mathcal{F} \Delta\psi/RT)}{1 - \exp(z_i\, \mathcal{F} \Delta\psi/RT)}\right] \qquad (2.13)$$

where $P_i = D_i \beta_i / \Delta x$.

Equation (2.13) is often referred to as the Goldman-Hodgkin-Katz equation and has been widely employed for the description of ion transport across biological membranes. However, it should be stressed that, strictly speaking, this equation only applies when the electrical potential profile across the membrane is linear and when the partition coefficients at the two interfaces are concentration independent. These conditions are most likely to be approached by thin membranes in which the density of fixed charges is low. On the other hand, this treatment is not generally valid for membranes (or pathways) that are thick and/or are characterized by a high fixed charge density. MacGillivray and Hare (1969) have shown that in order for the constant field assumption to be valid, a dimensionless value

$$\alpha^2 = (k_0 kRT/\mathcal{F}^2 \overline{X}\, \Delta x^2)$$

must be large; k_0 is the permittivity of free space, k is the dielectric constant of the barrier, and $\overline{X}$ is the concentration of fixed charge in the barrier. Pursuing this analysis, it can be shown that for a 100 Å-thick barrier with a dielectric constant of 9, the constant-field assumption will be valid (within experimental error) if $\overline{X} \leqslant 3 \times 10^{-4} M$.

Several other approaches have been taken toward the integration of the Nernst–Planck equation. These do not appear to offer any a priori advantages over the constant field assumption; the solutions are more complex and often unwieldy and, as a result, are encountered infrequently. These approaches are discussed briefly in Section 3.4 and are described in detail in Helfferich, 1962; Lakshminarayanaiah, 1969; MacInnes, 1961; and Sten-Knudsen, 1978.

2.3. Implications of the Goldman–Hodgkin–Katz (GHK) flux equation

It is of interest to explore some of the implications of the GHK equation with the aid of Figure 2.3, where we have plotted J_i versus $\Delta\psi$ when $c_i^o = 100$ mM, $P_i = 0.01$ cm/h, $z_i = +1$, and $c_i^i = 10$, 50, or 100 mM; J_i is expressed in microequivalents per centimeter squared, hour. When $J_i > 0$, the net flow is directed from compartment o to compartment i; when $J_i < 0$, the net flow is in the opposite direction.

(a) Clearly, when $c_i^o \neq c_i^i$, the relation between J_i and $\Delta\psi$ is curvilinear and the slope of the curve when $J_i < 0$ is less than the slope when $J_i > 0$. Thus, even when P_i is constant, the membrane offers greater *resistance* to the flow of this monovalent cation (per given change in $\Delta\psi$) when it is directed from a lower to a higher concentration (compartment i to compartment o) than when it takes place in the opposite direction. This asymmetric behavior is referred to as ("Goldman-type") *rectification*. The physical basis of this phenomenon will be discussed in Section 7.1, but it should be reemphasized at this point that it is not the result of a voltage-dependent change in the permeability of the barrier.

(b) According to equation (2.13), when $J_i = 0$, the intersections on the abscissa correspond to the values

$$\Delta\psi = (RT/\mathcal{F}) \ln (c_i^o/c_i^i)$$

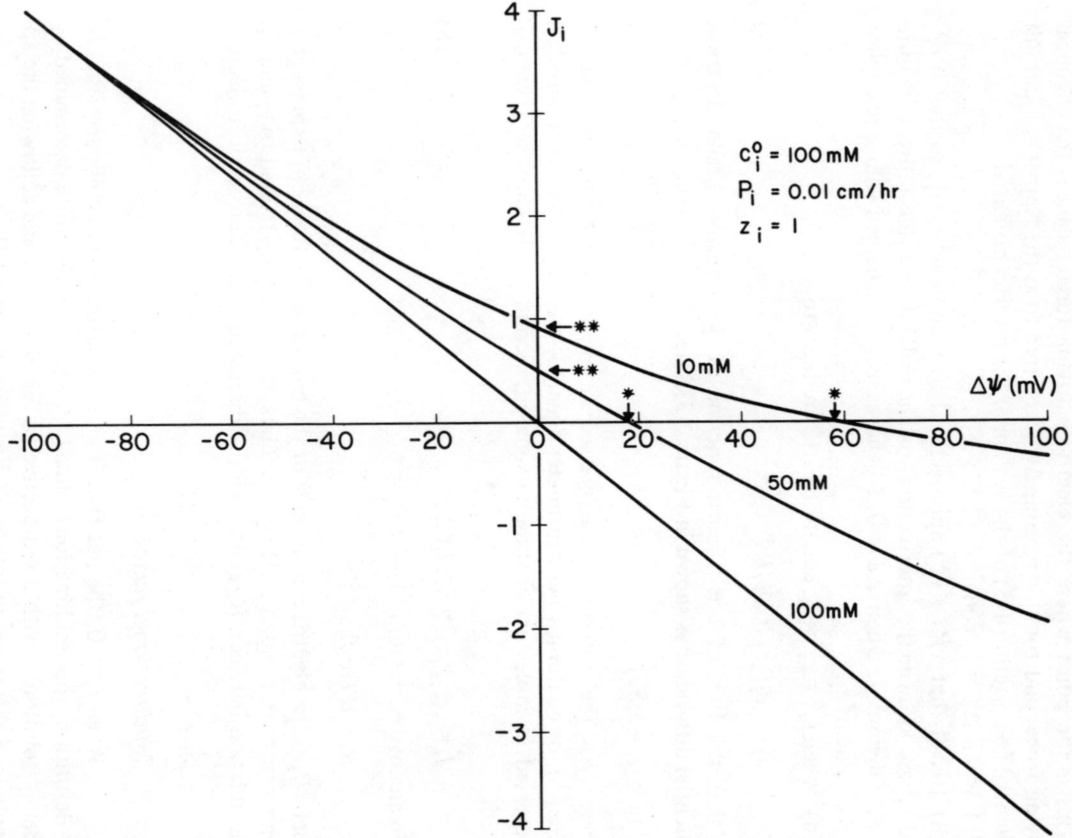

Figure 2.3. The relations between J_i and $\Delta\psi$ predicted by the GHK equation (2.13) for a monovalent cation when $c_i{}^o$ and P_i are maintained constant and $c_i{}^i$ is varied. The arrows with a single asterisk indicate the reversal potentials, with a double asterisk the points where $J_i = P_i \, \Delta c_i$.

This is the point where the electrical driving force balances the chemical driving force, and for this reason, $\Delta\psi$ is referred to as the Nernst equilibrium potential (see Section 1.5) of the system or the *reversal potential.*

(c) When $\Delta\psi = 0$, equation (2.13) reduces to $J_i = P_i(c_i^o - c_i^i)$ or equation (2.8) [recall that $x/(1 - e^x)$ approaches 1 as x approaches 0], so that if c_i^o and c_i^i are known, the intersections on the ordinate provide a direct measure of P_i. Obviously, when $\Delta\psi = 0$, i behaves as an uncharged solute regardless of the value of z_i.

(d) When $c_i^o = c_i^i = c_i$, equation (2.13) reduces to

$$J_i = -c_i P_i z_i \mathcal{F} \Delta\psi / RT \tag{2.14}$$

so that the flux of i is a linear function of the transmembrane electrical potential difference as shown in Figure 2.3. Since

$$I_i = z_i \mathcal{F} J_i \tag{2.15}$$

where I_i is the current of i expressed in amperes per centimeter squared (A/cm^2), or coulombs per centimeter squared, second (C/cm^2 sec), and J_i is expressed in moles per centimeter squared, second (moles/cm^2 sec), we can write

$$I_i = -(c_i P_i z_i^2 \mathcal{F}^2 / RT)\, \Delta\psi \tag{2.16}$$

In analogy with Ohm's law, $I_i R_i = \Delta\psi$, we see that

$$R_i = RT / c_i P_i z_i^2 \mathcal{F}^2 \tag{2.17}$$

where R_i is the electrical resistance of the barrier to the flow of i expressed in ohms centimeter squared ($\Omega \cdot cm^2$). Thus, R_i is inversely related to c_i and P_i. The relation between R_i and c_i will be discussed in greater detail in Section 7.1.

Unidirectional fluxes

When $c_i^i = 0$, the net flux from compartment o to compartment i is, by definition, the unidirectional flux J_i^{oi}, which is most often determined by labeling solution o with a radioactive tracer of solute i and following the appearance of this tracer in solution i. From equation (2.13)

$$(J_i)_{c_i^i = 0} = J_i^{oi} = P_i c_i^o \left[\frac{z_i \mathcal{F} \Delta\psi / RT}{1 - \exp(z_i \mathcal{F} \Delta\psi / RT)} \right] \tag{2.18}$$

When $\Delta\psi \to 0$, $(z_i \mathcal{F}\Delta\psi/RT)/[1 - \exp(z_i \mathcal{F}\Delta\psi/RT)] \to 1$ so that

$${}_0J_i^{oi} = P_i c_i^o \tag{2.19}$$

where ${}_0J_i^{oi}$ is the unidirectional flux of i when $\Delta\psi = 0$; this statement is analogous to equation (2.8) for the diffusion of an uncharged solute when $c_i^i = 0$.

Further, it can be shown that when $\Delta\psi < 25$ mV, $(z_i \mathcal{F}\Delta\psi/RT)/[1 - \exp(z_i \mathcal{F}\Delta\psi/RT)] \cong \exp(-z_i \mathcal{F}\Delta\psi/2RT)$, so that for small values of $\Delta\psi$

$$J_i^{oi} \cong P_i c_i^o \exp(-z_i \mathcal{F}\Delta\psi/2RT) \tag{2.20}$$

Similarly, when $c_i^o = 0$, from equation (2.13)

$$J_i^{io} = P_i c_i^i \left[\frac{(-z_i \mathcal{F}\Delta\psi/RT)\exp(z_i \mathcal{F}\Delta\psi/RT)}{1 - \exp(z_i \mathcal{F}\Delta\psi/RT)}\right]$$

or

$$J_i^{io} = P_i c_i^i \left[\frac{(-z_i \mathcal{F}\Delta\psi/RT)}{\exp(-z_i \mathcal{F}\Delta\psi/RT) - 1}\right]$$

and when $\Delta\psi < 25$ mV

$$J_i^{io} \cong P_i c_i^i \exp(z_i \mathcal{F}\Delta\psi/2RT)$$

Combining this last equation with equation (2.20), we obtain

$$J_i = J_i^{oi} - J_i^{io} = P_i[c_i^o \exp(-z_i \mathcal{F}\Delta\psi/2RT) - c_i^i \exp(z_i \mathcal{F}\Delta\psi/2RT)] \tag{2.21}$$

which is a reasonable approximation when $\Delta\psi < 25$ mV and is in error by only 15 percent when $\Delta\psi = \pm 50$ mV (see notes).[6]

The independence principle

An important implication of the GHK flux equation is the so-called *independence principle*, which asserts that the movement of a single ion at a point within the membrane is influenced only by the electrochemical gradient at that point and is *independent* of the presence of other ions of the same or different species. Three consequences of the GHK equation that illustrate the independence principle are as follows.

First, from equation (2.13), we see that

$$\frac{J_i}{J_i'} = \left[\frac{c_i^o - c_i^i \exp(z_i \mathcal{F}\Delta\psi/RT)}{c_i^{o\prime} - c_i^{i\prime} \exp(z_i \mathcal{F}\Delta\psi/RT)}\right]$$

where $c_i^{o\prime}$ and $c_i^{i\prime}$ represent a new set of concentrations that differ from c_i^o and c_i^i. Thus, if we double the concentrations of i on both sides of the membrane, holding $\Delta\psi$ constant, the net flux will double.

Second, for two species i and j, equation (2.13) predicts that

$$\frac{J_i}{J_j} = \left[\frac{P_i z_i [c_i^o - c_i^i \exp(z_i \mathcal{F}\Delta\psi/RT)]}{P_j z_j [c_j^o - c_j^i \exp(z_j \mathcal{F}\Delta\psi/RT)]}\right]$$

Thus, when $z_i = z_j$, $c_i^o = c_j^o$ and $c_i^i = c_j^i$

$$(J_i/J_j) = (P_i/P_j)$$

Finally, equations (2.18), (2.20) and (2.21) imply that the unidirectional fluxes of i are separable and influenced independently. Thus, J_i^{io} is independent of c_i^o and J_i^{oi} is independent of c_i^i. Or, stated in another way, there are no *transeffects* of concentration on a unidirectional flux.

As we will see below, there are a number of diffusional systems that exhibit significant departures from the independence relation and, thus, cannot be described using the GHK equation or any integrated form of the Nernst-Planck equation.

2.4. The irreversible thermodynamic approach

For an entirely phenomenologic description of diffusion of a solute across a barrier from one solution to another, it is often more convenient to define the driving force as the electrochemical potential difference ($\Delta\tilde{\mu}_i$) across the barrier rather than the negative of the electrochemical potential gradient ($-d\tilde{\mu}_i/dx$). This approach assumes that the electrochemical potential of i just within the membrane at the interface is equal to that in the adjacent solution, that is, $\bar{\tilde{\mu}}_i^o = \tilde{\mu}_i^o$ and $\bar{\tilde{\mu}}_i^i = \tilde{\mu}_i^i$. In essence, this implies that diffusion through the membrane is sufficiently slow so that the distribution of i across the interfaces (a rapid process) is essentially at equilibrium. Although $d\tilde{\mu}_i/dx$ and $\Delta\tilde{\mu}_i$ have different dimensions, it can be shown that, given these assumptions,

$$T(d_iS/dt) \equiv \sum_i J_i(-d\tilde{\mu}_i/dx) \equiv \sum_i J_i(\Delta\tilde{\mu}_i)$$

It should be noted that the "switch" in sign is not important since the difference in $\tilde{\mu}_i$ can be chosen so that it is positive and $J_i(\Delta\tilde{\mu}_i) > 0$ (Katchalsky and Curran, 1965, p. 115).

Using this approach and recalling that $\Delta\psi = \psi^i - \psi^o$

$$J_i = L_{ii}(\Delta\tilde{\mu}_i) = L_{ii}(RT\Delta \ln c_i - z_i \mathcal{F}\Delta\psi) \tag{2.22}$$

When $\Delta c_i = c_i^o - c_i^i$ is *small*

$$J_i = L_{ii}[RT(\Delta c_i/\dot{c}_i) - z_i \mathcal{F}\Delta\psi] \tag{2.23}$$

where $\dot{c}_i$ is an "average" or "mean" concentration given by

$$\dot{c}_i = \frac{c_i^o - c_i^i}{\ln(c_i^o/c_i^i)} \cong \frac{(c_i^o + c_i^i)}{2} \tag{2.24}$$

We now define $L_{ii} = \dot{c}_i P_i/RT$, so that

$$J_i = P_i \Delta c_i - P_i \dot{c}_i z_i \mathcal{F}\Delta\psi/RT \tag{2.25}$$

Thus, when Δc_i is small, J_i is the *sum* of a flow caused by the concentration difference *alone* (as if the solute is uncharged) and a flow that is a linear function of $\Delta\psi$.

Clearly, when $z_i = 0, J_i = P_i \Delta c_i$ [equation (2.8)], and when $c_i^o = c_i^i$ so that $\Delta c_i = 0$ and $\dot{c}_i = c_i$

$$-J_i = c_i P_i z_i \mathcal{F}\Delta\psi/RT \tag{2.26}$$

[see equation (2.14)].

2.5. The flux-ratio equation

In spite of the fact that the Nernst–Planck equation cannot be solved explicitly for J_i without additional assumptions, an extremely important relation between the bidirectional ("tracer") fluxes of i across a membrane (J_i^{oi} and J_i^{io}) can be derived from equation (2.4); this relation was derived by Behn (1897) and Teorell (1949, 1953) but was rederived and applied by Ussing (1949) and generally bears the eponym "the Ussing (flux-ratio) equation."

Assume that j, a tracer of solute i, is placed in solution o and that its concentration (activity) is c_j^o. Since $c_j^i = 0$, the (unidirectional) flux of j from o to i according to equation (2.11) is

$$J_j^{oi} \int_o^i \exp\left(\frac{z_i \mathcal{F}\bar{\psi}}{RT}\right) dx = D_i \bar{c}_j^o \tag{2.27}$$

If k, another tracer of solute i, is placed in solution i so that its concentration (activity) is c_k^i, then since $c_k^o = 0$, the (unidirectional) flux of k from i to o is given by

$$J_k^{io} \int_o^i \exp\left(\frac{z_i \mathcal{F}\bar{\psi}}{RT}\right) dx = D_i \bar{c}_k^i \left[\exp\left(\frac{z_i \mathcal{F}\Delta\bar{\psi}}{RT}\right)\right] \tag{2.28}$$

Note that equations (2.27) and (2.28) assume that $D_i = D_j = D_k$, which is implicit in the notion that j and k are "perfect tracers" for i.

Dividing equation (2.27) by equation (2.28), we obtain

$$\left[\frac{J_j^{oi}}{J_k^{io}}\right] = \left[\frac{\bar{c}_j^o}{\bar{c}_k^i}\right]\left[\exp\left(\frac{-z_i \mathcal{F}\Delta\bar{\psi}}{RT}\right)\right] \tag{2.29}$$

If the diffusion of j and k through the membrane is slow compared to the rate at which these tracers distribute across the membrane-solution interfaces, so that $\tilde{\mu}_j^o \cong \bar{\tilde{\mu}}_j^o$ and $\tilde{\mu}_k^i \cong \bar{\tilde{\mu}}_k^i$ then, as discussed in Section 1.5

$$\bar{c}_j^o = c_j^o \exp\left[\frac{-z_i \mathcal{F}(\bar{\psi}^o - \psi^o)}{RT}\right]$$

and

$$\bar{c}_k^i = c_k^i \exp\left[\frac{-z_i \mathcal{F}(\bar{\psi}^i - \psi^i)}{RT}\right]$$

Substituting these relations into equation (2.29), we obtain

$$\left[\frac{J_j^{oi}}{J_k^{io}}\right] = \left[\frac{c_j^o}{c_k^i}\right]\left[\exp\left(\frac{-z_i \mathcal{F}\Delta\psi}{RT}\right)\right] \tag{2.30}$$

If j and k are *perfect* tracers for the movements of i, then $(J_j^{oi}/c_j^o) = (J_i^{oi}/c_i^o)$ and $(J_k^{io}/c_k^i) = (J_i^{io}/c_i^i)$ so that

$$\left[\frac{J_i^{oi}}{J_i^{io}}\right] = \left[\frac{c_i^o}{c_i^i}\right]\left\{\exp\left[\frac{z_i\mathcal{F}(\psi^o - \psi^i)}{RT}\right]\right\} \tag{2.31}$$

Clearly, when $J_i^{oi} = J_i^{io}$, $J_i = 0$ and

$$\left[\frac{c_i^o}{c_i^i}\right] = \exp\left(\frac{z_i \mathcal{F}\Delta\psi}{RT}\right) \tag{2.32}$$

which, as pointed out before, is the *Nernst equilibrium equation* or the relation between the electrical potential difference and the concentration ratio when there is no net ion flow.

The original assumptions underlying the flux-ratio equation (2.31) are: (a) the solutions surrounding the membrane are perfectly stirred and the solute i (or its tracers j and k) are distributed at thermodynamic equilibrium across the two solution-membrane interfaces; (b) the chemical state of i within the membrane is identical with that in the surrounding solutions (i.e., there are no chemical reactions, associations, complex formation, etc. within the membrane); (c) the unidirectional fluxes J_i^{oi} and J_i^{io} traverse the membrane through pathways that have identical properties [if not, the division of equation (2.27) by equation (2.28) to obtain equation (2.29) would not be valid]; and (d) the unidirectional fluxes of the tracers j and/or k are not affected by interactions with the flow of the parent isotope i or the flows of other solutes or solvent (Ussing, 1949). Recently, more general derivations have been described by Hoshiko and Lindley (1967), Kedem and Essig (1965), and Dawson (1977). Further, the flux-ratio equation has been shown to be valid for the cases of substances formed or consumed during passage across the membrane (Ussing, 1952), for membranes that are inhomogeneous in all three dimensions (Schwartz, 1971*b*), and under non-steady-state conditions (Ussing, 1972).

The power of the flux-ratio equation resides in the fact that it does not depend upon a detailed knowledge of the properties of the membrane, and, in particular, no assumption need be made regarding the electrical potential profile within the membrane. Thus, in obtaining the ratio of fluxes, both D_i and the integral on the left-hand sides of equations (2.27) and (2.28) cancel and the final result can be expressed entirely in terms of readily measurable external parameters. For this reason, the flux-ratio equation has provided a valuable criterion for determining whether the bidirectional and, thus, the net flux of an ion across a barrier can be attributed entirely to electrical and chemical potential differences. If the bidirectional fluxes of an ion conform to equation (2.31) over a wide range of concentrations and electrical potential differences, we can confidently conclude that transport is caused by simple diffusion. On the other hand, as will be discussed in Section 5.2, the converse is not true; that is, failure of the movements of an ion to conform to the flux-ratio equation does not exclude the possibility that its net movement is energized solely by differences in chemical and/or electrical potentials.

Finally, it should be noted that equation (2.31) can be derived from equations (2.18), (2.20) and (2.21), but it should be stressed that these equations rest on the validity of the constant-field assumption.

2.6. Discontinuous diffusion

The formal descriptions of diffusional processes derived by integrating the Nernst-Planck equation, which necessarily assume that $(d\tilde{\mu}_i/dx)$ is a continuous function and which embody the independence principle, do not adequately describe systems in which the diffusing particle interacts with specific sites along its path.[7] As we will see, such systems can display saturation, transeffects, and competitive behavior that cannot be described by any treatment in which P_i, the permeability coefficient, is a unique property of the diffusing species and the barrier and is not influenced by the concentration of the diffusing species or the presence of other permeant ions.

An approach that has proved useful in analyzing such systems, based on rate theory (Glasstone et al., 1941), was originally proposed by Eyring, Lumry, and Woodbury (1949) and has been recently elegantly extended by Heckmann (1972), Läuger (1973), Hille (1975), Sandblom, Eisenman, and Neher (1977), and others (cf. Hille and Schwarz, 1978). This approach can become mathematically formidable and has not been completely developed for complex systems. For these reasons, we will simply try to illustrate the underlying principles employing relatively simple examples.

The fundamental assumption is that the membrane can be viewed as a series of potential energy barriers, depicted as a series of peaks and valleys, that a particle must cross in order to pass from the outer (o) to the inner (i) solution. Every hindrance to advancement (electrostatic or steric) is represented by a peak that denotes the minimum amount of energy the particle must possess to traverse this obstacle. Conversely, every favorable region for residence is represented by a valley from which the particle must escape in order to advance further. The rate constant for advancement is derived from statistical mechanical reasoning and, in particular, from the "Boltzmann distribution." Thus, the rate constant for movement of a particle from valley j to valley k, over the peak jk is given by

$$k_{jk} = K(kT/h)\exp(-\Delta G_{jk}/RT) \tag{2.33}$$

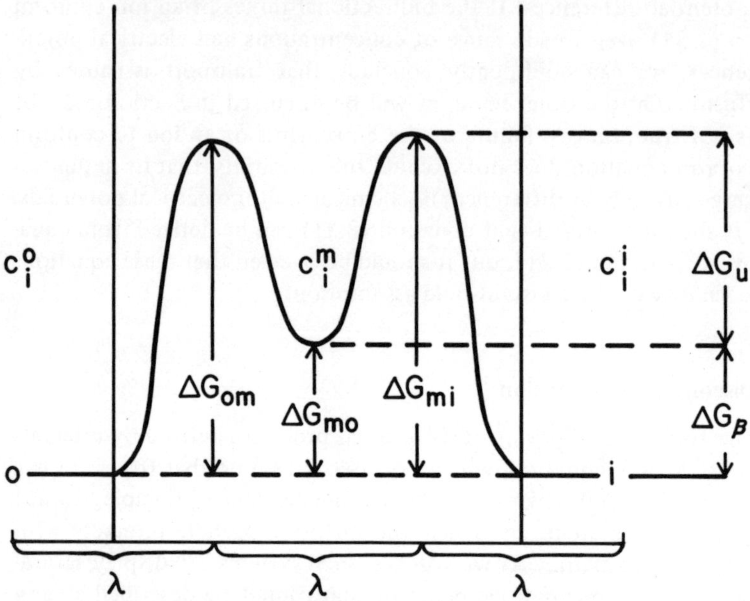

Figure 2.4. A simple, double energy-barrier model for diffusion across a membrane.

where ΔG_{jk} is the height of the peak from the valley j, k is the Boltzmann constant ($k = R/N$ where N is Avogadro's number) and h is the Planck constant. K is referred to as the *transmission coefficient* and represents the probability that a particle that successfully reaches the peak will advance to valley k rather than "slide back" into valley j; K is generally assigned a value of unity.

The term multiplied by the transmission coefficient is simply the probability that a given particle will possess at least the amount of free energy ΔG_{jk} necessary to reach the peak.

We now illustrate the application of this reasoning to the simple two-barrier representation of a membrane illustrated in Figure 2.4, where the distance between energy peaks is designated as λ. If $c_i{}^o$ is the concentration of diffusing particles in the outer solution and $c_i{}^m$ is the concentration in the membrane, then the velocity of flow from the outer solution into the membrane is

$$v_i{}^{om} = k_{om}\, c_i{}^o \lambda - k_{mo}\, c_i{}^m \lambda \qquad (2.34)$$

If the area of the membrane is 1 cm^2 and λ is in centimeters, then $c_i{}^o\lambda$ is the number of molecules in the lamina with thickness λ preceding the first energy peak; k_{om} has units of seconds^{-1} so that $v_i{}^{om}$ is expressed in molecules per centimeter squared second.

Similarly, we can write for the velocity of exit from the membrane into solution i

$$v_i^{mi} = k_{mi} c_i^m \lambda - k_{im} c_i^i \lambda \tag{2.35}$$

When the diffusion of i from o to i achieves a steady state, c_i^m is constant and $v_i^{om} = v_i^{mi} = J_i$. Also, if the barriers have the same amplitudes, $k_{om} = k_{im}$ and $k_{mo} = k_{mi}$ so that solving equations (2.34) and (2.35), we obtain

$$J_i = (k_{om} \lambda/2)(c_i^o - c_i^i) \tag{2.36}$$

For an uncharged molecule, equation (2.36) has the same form as equation (2.8), so that we may write $P_i = (k_{om}\lambda/2)$. Since $P_i = D_i\beta_i/\Delta x$ and $\Delta x = 2\lambda$, it follows that $D_i\beta_i = k_{om}\lambda^2$.

From equation (2.33), we see that $k_{om} = (kT/h)\exp(-\Delta G_{om}/RT)$. Further, we can divide ΔG_{om} into the difference in free energy of i between the outer phase and the valley in the membrane phase, ΔG_β, and the difference between the valley and the internal peaks, ΔG_u. Thus, we can write

$$D_i\beta_i = \lambda^2 (kT/h) \exp(-\Delta G_\beta/RT) \exp(-\Delta G_u/RT) \tag{2.37}$$

Clearly, the partition coefficient β_i can be identified with the expression $\exp(-\Delta G_\beta/RT)$.[5] Similarly, the diffusion coefficient can be identified with the expression $\lambda^2 (kT/h) \exp(-\Delta G_u/RT)$, where the exponential term represents the effect that energy barriers within the membrane have on the mobility of the diffusing molecule.

Several points can be generalized from this simple exercise. First, unlike the approaches that involve integration of the Nernst–Planck equation where u_i or D_i and β_i are phenomenologic coefficients, the rate theory analysis, in principle, outlines a way in which these intramembrane properties can be expressed in terms of the physical and chemical structural details of the diffusion pathway.

Second, as pointed out by Eyring, Lumry, and Woodbury (1949), regardless of the complexity of the system, the only quantities needed to define P_i are the amplitudes of the internal peaks relative to the external solutions; the depths of the valleys between peaks do not appear in the final expression. Thus, the factors affecting partitioning (binding) and mobility appear as a product [equation (2.37)] and need not be dissociated in order to explicitly define permeability.

In short, given sufficient information regarding the microscopic details of the barrier, this analysis offers an approach to the description of diffusional flows at the elemental level.

We now consider the diffusion of a charged species through a narrow pore that contains a binding site X with the restriction that, in order to traverse the pore, i must bind to X and that X can only be occupied by one i at a time. The energy profile of this barrier can be represented as shown in Figure 2.5, where Δx is the thickness of the membrane and λ designates the posi-

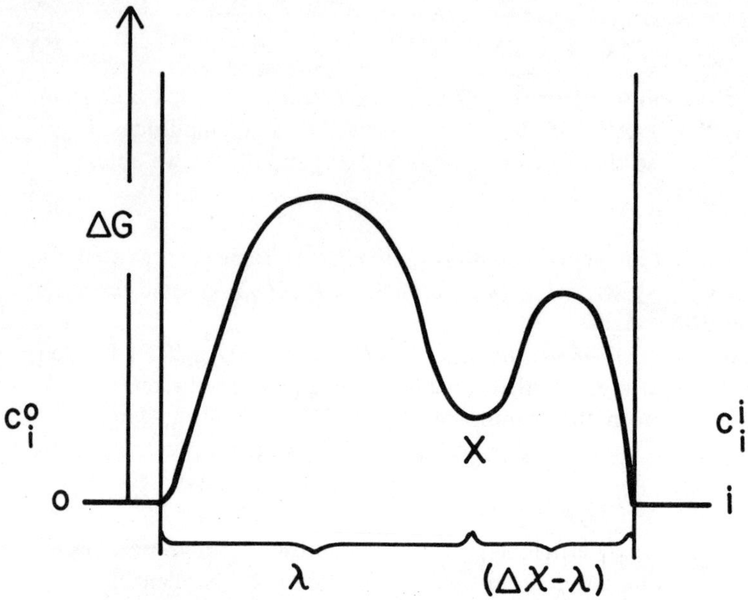

Figure 2.5. Energy profile for ionic diffusion through a pore that contains a fixed site X capable of binding the ion i.

tion of X; ΔG represents the free energy profile caused by both an external electrical potential ($\Delta\psi$) and internal barriers. As discussed by Hille (1975), this system can be described by the equations

$$c_i^o + [X] \underset{k_{xo}^*}{\overset{k_{ox}^*}{\rightleftharpoons}} [c_i X]$$

and (2.38)

$$[c_i X] \underset{k_{ix}^*}{\overset{k_{xi}^*}{\rightleftharpoons}} c_i^i + [X]$$

where the brackets designate the concentrations of the free and complexed forms of X.

Equations (2.38) are strictly analogous to the equations of chemical kinetics describing the reversible reaction between a substrate and an enzyme leading to the formation of a product. The rate constants are given by the following relations:

$$k_{ox}^* = k_{ox} \exp\left[-z\mathcal{F}(1 - \alpha)\,\psi^{ox}/RT\right]$$

$$k_{xo}^* = k_{xo} \exp\left[z\mathcal{F}\alpha\psi^{ox}/RT\right]$$

$$k_{xi}^* = k_{xi} \exp\left[-z\mathcal{F}(1 - \beta)\,\psi^{xi}/RT\right] \qquad (2.39)$$

and

$$k_{ix}{}^{*} = k_{ix} \exp\,(z\,\mathcal{F}\beta\,\psi^{xi}/RT)$$

where ψ^{ox} is the electrical potential at X with respect to the outer solution so that $\Delta\psi = \psi^i - \psi^o = \psi^{ox} + \psi^{xi}$. The k's (without asterisks) are independent of the external field but are dependent on voltage-independent internal energy barriers as described by equation (2.33). The values of α and β are dependent upon the electrical potential profile between o and λ and between λ and i; if these profiles are symmetrical about the midpoints the rate constants will be equally and oppositely affected by the applied potential and $\alpha = \beta = 0.5$ (Hall, et al., 1973; Stein, 1977; Glasstone et al., 1941).

From equation (2.38) the net flux of i from compartment o to compartment i under steady-state conditions is given by

$$J_i = k_{ox}{}^{*} c_i{}^{o}\,[X] - k_{xo}{}^{*}\,[c_i X] = k_{xi}{}^{*}\,[c_i X] - k_{ix}{}^{*} c_i{}^{i}\,[X]$$

Further, the total concentration of binding sites per unit area $[X]_t$ is equal to the concentrations of those sites occupied by i, $[c_i X]$, and those which remain unoccupied $[X]$, so that

$$[X]_t = [X] + [c_i X]$$

Solving these equations for J_i, we obtain

$$J_i = [X]_t \left(\frac{k_{ox}{}^{*} k_{xi}{}^{*} c_i{}^{o} - k_{ix}{}^{*} k_{xo}{}^{*} c_i{}^{i}}{k_{xo}{}^{*} + k_{xi}{}^{*} + k_{ix}{}^{*} c_i{}^{i} + k_{ox}{}^{*} c_i{}^{o}} \right) \tag{2.40}$$

From equations (2.39) and (2.40) it can be readily shown that $J_i = 0$ when

$$\left(\frac{c_i{}^{o}}{c_i{}^{i}} \right) = \left(\frac{k_{ix}{}^{*} k_{xo}{}^{*}}{k_{ox}{}^{*} k_{xi}{}^{*}} \right) = \left(\frac{k_{ix}\, k_{xo}}{k_{ox}\, k_{xi}} \right) \exp \left(\frac{z_i\, \mathcal{F} \Delta\psi}{RT} \right) \tag{2.41}$$

Equation (2.41) conforms to the Nernst equilibrium expression (2.32) only when $(k_{ix}\, k_{xo}/k_{ox}\, k_{xi}) = 1$, which is one of the constraints on the rate constants in a strictly diffusional system.

From equation (2.40) it is clear that the net flux from compartment o to compartment i when $c_i{}^{i} = 0$ is given by

$$J_i = [X]_t\, \frac{(k_{ox}{}^{*} k_{xi}{}^{*} c_i{}^{o})}{(k_{xo}{}^{*} + k_{xi}{}^{*} + k_{ox}{}^{*} c_i{}^{o})} \tag{2.42}$$

Clearly, equation (2.42) has the form of the familiar Michaelis-Menten equation of enzyme kinetics. J_i saturates with increasing $c_i{}^{o}$ (at constant $\Delta\psi$); the maximum influx under these conditions is $[X]_t\, k_{xi}{}^{*}$ and a half-maximal influx is achieved when $c_i{}^{o} = (k_{xo}{}^{*} + k_{xi}{}^{*})/k_{ox}{}^{*}$.

Conversely, when $c_i{}^{o} = 0$, the net flux from compartment i to compartment o is given by

$$J_i = [X]_t\, \frac{k_{ix}{}^{*} k_{xo}{}^{*} c_i{}^{i}}{k_{xo}{}^{*} + k_{xi}{}^{*} + k_{ix}{}^{*} c_i{}^{i}} \tag{2.43}$$

so that the maximum outflux is $[X]_t k_{xo}^*$ and the half-maximum outflux is achieved when $c_i^i = (k_{xo}^* + k_{xi}^*)/k_{ix}^*$.

Equations (2.42) and (2.43) disclose an extremely interesting property of this system; namely, the kinetic properties that describe the relation between J_i and c_i^o when $c_i^i = 0$ differ from those that describe the relation between J_i and c_i^i when $c_i^o = 0$. Even when $\Delta\psi = 0$, the membrane will behave symmetrically only when $k_{xo}^* = k_{xi}^*$ (i.e., the maximum fluxes in either direction are equal) *and* $k_{ix}^* = k_{ox}^*$ (i.e., the concentrations needed to achieve half-maximum fluxes are equal). Otherwise, the membrane will display rectification; that is, for a given value of c_i the flow will differ when i is in compartment o alone from that observed when i is in compartment i alone.

Another interesting feature of this system emerges when we consider the bidirectional fluxes determined with tracers. It can be readily shown that if a perfect tracer for i, k, is placed in compartment o, the tracer flux is given by

$$J_k^{oi} = [X]_t \frac{k_{ox}^* k_{xi}^* c_k^o}{k_{xo}^* + k_{xi}^* + k_{ox}^* c_i^o + k_{ix}^* c_i^i}$$

If k is a perfect tracer for i, it follows that $(J_k^{oi}/c_k^o) = (J_i^{oi}/c_i^o)$ so that

$$J_i^{oi} = [X]_t \frac{k_{ox}^* k_{xi}^* c_i^i}{k_{xo}^* + k_{xi}^* + k_{ox}^* c_i^o + k_{ix}^* c_i^i} \tag{2.44}$$

Similarly, the unidirectional flux from compartment i to compartment o is given by

$$J_i^{io} = [X]_t \frac{k_{ix}^* k_{xo}^* c_i^i}{k_{xo}^* + k_{xi}^* + k_{ox}^* c_i^o + k_{ix}^* c_i^i} \tag{2.45}$$

Thus, the unidirectional fluxes exhibit transeffects; that is, the unidirectional flux from o to i is affected by c_i^i and the unidirectional flux from i to o is affected by c_i^o.

Finally, from equations (2.44) and (2.45) it is clear that the flux ratio is given by

$$\left(\frac{J_i^{oi}}{J_i^{io}}\right) = \left(\frac{c_i^o}{c_i^i}\right)\left(\frac{k_{ox}^* k_{xi}^*}{k_{ix}^* k_{xo}^*}\right)$$

which from equation (2.41) reduces to

$$\left(\frac{J_i^{oi}}{J_i^{io}}\right) = \left(\frac{c_i^o}{c_i^i}\right) \exp\left(\frac{-z_i \mathcal{F}\Delta\psi}{RT}\right)$$

or the Ussing flux-ratio equation.

The behavior of this very simple system obviously deviates significantly from those which conform to the independence principle, inasmuch as there is no unique permeability coefficient P_i that relates the net flux or unidirectional fluxes to differences in concentration and/or electrical potential. In contrast with systems described by the GHK equation (2.13), where P_i is independent of c_i and $\Delta\psi$, in this system P_i is explicitly dependent upon these parameters. In addition, these systems display rectification (even when $\Delta\psi = 0$) and transeffects. Finally, although this system conforms to the Ussing flux-ratio equation, as will be discussed in Section 5.2, pores involving multiple binding sites that permit only "single-file diffusion" do not conform to this relation.

Finally, as discussed by Hille (1975), and others (Heckman, 1972), the simple system illustrated in Figure 2.5 and described by equations (2.38) can exhibit competitive interactions that will influence the flows of i when another species with an affinity for X is present in the outer and/or inner solution; this should be intuitively obvious.

This approach has been successfully applied to the description of ion transport across artificial lipid membranes doped with "pore-forming" antibiotics. These pores appear to possess discrete sites with which a traversing ion interacts, and these sites appear to be separated by distances that are large compared to the size of the ion. Ionic currents through such systems exhibit saturation and are subject to competitive inhibition and, thus, do not conform to the independence principle. However, it should be noted that if we assume that the membrane is comprised of an infinite number of energy barriers (i.e., approaching a continuum), the expressions resulting from the rate theory analysis resemble those derivable from the Nernst–Planck equation; and the GHK flux equation can be derived if, in addition, we assume that all of the barriers are of equal height (Hille, 1975).

Another, but similar, approach to systems in which there are discontinuities in the chemical potential of the diffusing species i has been described by Hill and Kedem (1966), who view the membrane as a lattice made up of independent units each of which can exist in a finite number of discrete states. In a subsequent paper, Essig, Kedem, and Hill (1966) examine net flows and unidirectional (tracer) flows in such lattice models and explore the conditions under which the flux-ratio equation (2.31) is obeyed.

2.7. A few words about permeability coefficients

One serious deficiency of all approaches toward a formal description of solute transport across membranes is that there is no generally applicable way to calculate P_i from a knowledge of the properties of i and those of the barrier. The P_i that emerges from the various integrated forms of the Nernst–Planck equation is strictly a secondarily derived, phenomenologic ("black box") coefficient that relates the observed flux to the driving force. The rate theory approach outlines, in principle, the energetic factors that determine P_i, but the application of this approach to complex membranes presents formidable if not insurmountable practical problems.

Limited success in deriving values of P_i for the very simple case of diffusion of small uncharged molecules through water-filled, right-cylindrical pores was first achieved by Pappenheimer, Renkin, and their collaborators (Pappenheimer, 1953). In analogy with equations (2.7) and (2.8), they wrote

$$J_i = D_{is} A_s \, \Delta c_i / \Delta x$$

where D_{is} is the diffusion coefficient of the solute in an aqueous solution (rather than in the membrane phase) and A_s lumps all of the membrane properties that influence entry into and movement through the pore. Thus, $P_i = D_{is} A_s / \Delta x$. Renkin (1954) assumed that the total restriction on diffusion through the pore, compared to unrestricted (free) aqueous diffusion, can be divided into (a) steric hindrance at the orifice of the pore, which influences the probability that a randomly moving particle will enter the pore (partitioning) and (b) hindrances to movement within the pore arising from frictional interactions with the wall (i.e., the difference between D_{is} and D_i). Pappenheimer, Renkin and Borrero (1951) suggested that the likelihood that a particle will enter a pore (i.e., the probability that the particle will not collide with the rim of the pore) is given by $[1 - (r_i/r_p)]^2$, where r_i is the molecular radius of the solute and r_p is the radius of the pore. The effect of frictional interactions within the pore was described by the Faxen equation, which was derived for the case of spheres moving through a fluid capillary assuming that the spheres are much larger than the fluid molecules (i.e., the fluid can be treated as a continuum). Using these assumptions, Renkin (1954) showed that

$$\frac{P_i}{D_{is}} = \frac{A_s}{\Delta x} = \frac{A_p}{A \, \Delta x} \cdot \left[1 - \left(\frac{r_i}{r_p}\right)\right]^2 \left[1 - 2.109 \left(\frac{r_i}{r_p}\right) + 2.09 \left(\frac{r_i}{r_p}\right)^3 - 0.95 \left(\frac{r_i}{r_p}\right)^5\right] \qquad (2.46)$$

where A_p/A is the fraction of the total membrane area A occupied by pore orifices A_p. Thus, if A, Δx, A_p, and r_p are known for a given membrane, P_i can be calculated from the knowledge of r_i and D_{is} (which can be determined experimentally).

The Renkin equation (2.46) has been verified in experiments using membranes that have uniform, well-defined thicknesses and that are penetrated by pores whose radii are known (Beck and Schultz, 1970; House, 1974). However, it should be stressed that for the case of biological membranes: (a) r_p and A_p cannot be determined directly; (b) the "pores" (if any) are almost certainly tortuous channels that cannot be characterized by a single radius; and (c) the pore dimensions are such that the molecules that are likely to enter are not very much larger than water, so that the assumptions underlying the use of the Faxen equation are unrealistic. Thus, while the application of the Renkin equation can provide an internally consistent body of results yielding "relative permeabilities," "equivalent pore radii," and so on, these deductions may not be very realistic.

At present there is no generally accepted approach to the calculation of the permeability coefficients of charged species from the properties of the ion and the barrier. Clearly, a formalism that would permit an independent calculation of P_i from first principles would provide an independent check of the validity of the flux equations employed to describe diffusional flows.

3 Diffusion potentials

In Chapter 2 we considered the effect of electrical potential differences on the diffusional flow of ions across a barrier without reference to the origin of these potentials. In the following sections we will examine how the diffusion of ions gives rise to electrical potential differences or *diffusion potentials*, starting with some simple, specific examples and concluding with more general treatments.

3.1. Diffusion potential between two solutions of the same salt

Assume that a homogeneous membrane separates two well-stirred solutions (o and i) of a salt that is completely dissociated into a cation c_+ and an anion c_-, where $z_+ = -z_-$. According to the Nernst–Planck equation,

$$J_+ = -\bar{c}_+ u_+ \left[RT \left(\frac{d \ln \bar{c}_+}{dx} \right) + z_+ \mathcal{F} \left(\frac{d\bar{\psi}}{dx} \right) \right] \tag{3.1}$$

and

$$J_- = -c_- u_- \left[RT \left(\frac{d \ln \bar{c}_-}{dx} \right) - z_+ \mathcal{F} \left(\frac{d\bar{\psi}}{dx} \right) \right] \tag{3.2}$$

Since bulk electroneutrality of each solution must be preserved,[1] it follows that $c_+{}^o = c_-{}^o = c_i{}^o$, $c_+{}^i = c_-{}^i = c_i{}^i$, and $J_+ = J_- = J_i$. Equating (3.1) and (3.2) and rearranging the terms yields

$$\frac{d\bar{\psi}}{dx} = -\left(\frac{u_+ - u_-}{u_+ + u_-} \right) \left(\frac{RT}{z_+ \mathcal{F}} \right) \left(\frac{d \ln \bar{c}_i}{dx} \right) \tag{3.3}$$

and integrating across the thickness of the membrane, we obtain

$$\Delta\bar{\psi} = \bar{\psi}^i - \bar{\psi}^o = -\left(\frac{u_+ - u_-}{u_+ + u_-} \right) \left(\frac{RT}{z_+ \mathcal{F}} \right) \ln \left(\frac{\bar{c}_i{}^i}{\bar{c}_i{}^o} \right)$$

Assuming that $\beta_i = \bar{c}_i{}^i / c_i{}^i = \bar{c}_i{}^o / c_i{}^o$, we obtain[2]

$$\Delta\psi = \psi^i - \psi^o = -\left(\frac{u_+ - u_-}{u_+ + u_-} \right) \left(\frac{RT}{z_+ \mathcal{F}} \right) \ln \left(\frac{c_i{}^i}{c_i{}^o} \right) \tag{3.4}$$

When $z_+ \neq -z_-$, it can be readily shown that

$$\Delta\psi = -\left(\frac{u_+ - u_-}{z_+u_+ + z_-u_-}\right)\left(\frac{RT}{\mathcal{F}}\right)\ln\left(\frac{c_i^{\,i}}{c_i^{\,o}}\right)$$

(Schwartz, 1971b).

Substituting equation (3.3) into equation (3.1) and combining terms yields

$$J_+ = -\left(\frac{2RTu_+u_-}{u_+ + u_-}\right)\frac{dc_i}{dx} \tag{3.5}$$

Since $J_+ = J_i$, equation (3.5) can be rewritten

$$J_i = -D_{\pm}(dc_i/dx)$$

where $D_{\pm}$, the diffusion coefficient of the *salt*, is simply

$$(2RTu_+u_-)/(u_+ + u_-) = RTu_{\pm}$$

The important principle to be gleaned from this simple example is central to the understanding of the origin of diffusion potentials in more complex systems. Namely, in the absence of an externally applied current, electroneutrality can only be preserved by the *equivalent* flow of anions and cations across the membrane. This so-called *zero current condition* is satisfied in this simple system by the constraint that $J_+ = J_- = J_i$ so that

$$I = I_+ + I_- = z_+\mathcal{F}J_+ + z_-\mathcal{F}J_- = 0 \tag{3.6}$$

If $u_+ \neq u_-$, equation (3.6) can only be satisfied if an electrical potential difference is generated *whose magnitude is proportional to the difference in mobilities and whose orientation is such as to retard the movement of the ion with the greater mobility and accelerate the movement of the ion with the lower mobility.* Thus, if the cation has a greater mobility than the anion, the dilute solution will be electrically positive with respect to the more concentrated solution and vice versa. The result is that the two oppositely charged species move with the same velocity, and the diffusion of a dissociated salt can be characterized by a single diffusion coefficient that is given by the *harmonic mean* of the two mobilities.

Stated in another way, if $u_+ = u_-$, then when the *same* force acts on both species, $v_+ = v_-$ and $J_+ = J_-$; thus, the only force needed is the concentration gradient (dc_i/dx) and, according to equation (3.4), $\Delta\psi = 0$. However, if $u_+ \neq u_-$, *another* force must be present that exerts *opposite* effects on the two species in order that $J_+ = J_-$; this force is the diffusion potential $\Delta\psi$.

We now examine the limiting condition where the membrane is ideally impermeable to one, but not both, of the charged species. According to equation (3.4), if $u_+ = 0$,

$$\Delta\psi = -\left(\frac{RT}{z_+\mathcal{F}}\right)\ln\left(\frac{c_i^{\,i}}{c_i^{\,o}}\right) \tag{3.7}$$

If $u_- = 0$, then $\Delta\psi$ has the same magnitude but the opposite orientation. However, if *both* ions cannot cross the membrane, then *neither* is permitted to cross, otherwise electroneutrality would be violated. Thus, if either $u_+ = 0$ *or* $u_- = 0$, $J_+ = J_- = 0$ and $D_\pm = 0$. Therefore, in this limiting condition, there is no movement and the system is in a state of equilibrium. The magnitude and orientation of $\Delta\psi$ precisely oppose the chemical potential difference of the permeant species so that there is no net driving force acting on that species and, hence, no flow.

More explicitly, if $u_+ = 0$, we can rearrange equation (3.7) as follows:

$$z_- \mathcal{F}\Delta\psi = -RT \ln(c_-^i/c_-^o)$$

or $\Delta\tilde{\mu}_- = RT \ln(c_+^i/c_+^o) + z_- \mathcal{F}\Delta\psi = 0$. Since $\Delta\tilde{\mu}_- = 0$, the anion is at equilibrium; $\Delta\psi$ is referred to as the *Nernst equilibrium potential* of the species and may be viewed as the electrical equivalent of the concentration difference (see Section 1.5).

We can now appreciate why interfacial potentials cancel [see derivation of equations (2.13) and (3.14)] if (a) translocation of an ion through the barrier is slow compared to the rate at which the ion can partition across the interface so that the interfacial distribution can be considered to be an equilibrium distribution and (b) the partition coefficient β_i is concentration independent. Clearly, under these conditions

$$(\overline{\psi}^o - \psi^o) = (RT/z_i\mathcal{F}) \ln \beta_i = (\overline{\psi}^i - \psi^i)$$

so that

$$(\psi^i - \psi^o) = (\overline{\psi}^i - \overline{\psi}^o) = \Delta\psi$$

3.2. The Goldman-Hodgkin-Katz equation

The problem becomes vastly more complex when we consider a system in which the membrane is surrounded by two solutions containing various anions and cations, not simply a single dissociated salt. The simplest and most frequently employed approach to such a system was originally derived by Goldman (1943) and later rederived and extended by Hodgkin and Katz (1949) for the restricted conditions where (a) anions and cations are univalent; (b) the permeability coefficient of each ion is constant, concentration independent and is given by $u_i\beta_i RT/\Delta x$; and (c) the electrical field across the membrane ($d\overline{\psi}/dx$) is constant. Under these conditions the flow of *each* cation is given by [equation (2.13)]

$$J_+ = -\frac{P_+ \mathcal{F}\Delta\psi}{RT}\left[\frac{c_+^o - c_+^i \exp(\mathcal{F}\Delta\psi/RT)}{1 - \exp(\mathcal{F}\Delta\psi/RT)}\right] \tag{3.8}$$

and that of *each* anion by

$$J_- = \frac{P_- \mathcal{F}\Delta\psi}{RT}\left[\frac{c_-^o - c_-^i \exp(-\mathcal{F}\,\Delta\psi/RT)}{1 - \exp(-\mathcal{F}\,\Delta\psi/RT)}\right] \tag{3.9}$$

Under "zero current conditions" (see Section 3.1)

$$I = \mathcal{F}\left(\sum_c J_+ - \sum_a J_-\right) = 0 \tag{3.10}$$

where $\Sigma_c J_+$ is the sum of the flows of all cations and $\Sigma_a J_-$ is the sum of the flows of all anions (recall that a flow is defined as positive when it is directed from o to i).

Combining equations (3.8), (3.9), and (3.10) and solving for $\Delta\psi$, we obtain the Goldman-Hodgkin-Katz (GHK) equation

$$\Delta\psi = \frac{RT}{\mathcal{F}} \ln\left(\frac{\sum_c P_+c_+^o + \sum_a P_-c_-^i}{\sum_c P_+c_+^i + \sum_a P_-c_-^o}\right) \tag{3.11}$$

In many biological systems where the predominant, permeant cations and anions are Na, K, and Cl, equation (3.11) is simply

$$\Delta\psi = \frac{RT}{\mathcal{F}} \ln\left(\frac{P_{Na}c_{Na}^o + P_K c_K^o + P_{Cl}c_{Cl}^i}{P_{Na}c_{Na}^i + P_K c_K^i + P_{Cl}c_{Cl}^o}\right) \tag{3.12}$$

Clearly, if an ion i is distributed at equilibrium across the membrane so that $\Delta\tilde{\mu}_i = 0$, then $J_i = 0$ and this ion will not participate in the generation of $\Delta\psi$. Thus, for example, if $\Delta\tilde{\mu}_{Cl} = 0$, equation (3.12) reduces to

$$\Delta\psi = \left(\frac{RT}{\mathcal{F}}\right) \ln\left(\frac{c_{Na}^o + \alpha c_K^o}{c_{Na}^i + \alpha c_K^i}\right) \tag{3.13}$$

where $\alpha = P_K/P_{Na}$.

Generalizations of the GHK equation

As noted, equation (3.11) was originally derived by assuming that the anions and cations are monovalent,[3] have constant permeability coefficients, and that the electrical potential profile across the membrane is linear (i.e., a constant electrical field). Subsequently, a number of generalizations have been introduced that relax some of these restrictions and extend the range of validity of expressions that have the form of the original GHK equation (3.11). Thus:

(a) If the membrane is permeable only to ions having the same sign and valence, then (Conti and Eisenman, 1965, 1966)

$$\Delta\psi = \frac{RT}{z_i \mathcal{F}} \ln\left(\frac{\sum_i P_i c_i^o}{\sum_i P_i c_i^i}\right) \tag{3.14}$$

Equation (3.14) can be derived directly from the Nernst-Planck equation (2.4) without any assumptions regarding the electrical potential profile within the membrane or interfacial potentials.

(b) Equation (3.11) is generally valid when the *total* concentration of univalent ions (or ions having the same absolute valence) is the same on both sides of the membrane; under this condition, it can be shown that the electrical field within the membrane is constant (Teorell, 1953; Finkelstein and Mauro, 1963; Sandblom and Eisenman, 1967).

(c) Equation (3.11) is valid if the individual permeability coefficients are not constant *providing that the ratios* of permeability coefficients, given by the ratio of the products of individual mobilities and partition coefficients, are constant (Sandblom and Eisenman, 1967); that is,

$$(P_i/P_j) = (u_i \beta_i / u_j \beta_j) = \text{constant for all } i \text{ and } j$$

(d) The "constant field" assumption is too restrictive. The only constraint on the electrical potential profile *within the membrane* required to derive equation (3.11) is that

$$\int_o^i \sinh(\mathcal{F}\overline{\psi}/RT)\, dx = 0$$

This constraint is consistent with the family of electrical potential profiles that are antisymmetric ("odd symmetry") about the midpoint of the membrane (Figure 3.1) (Mullins, 1961; Barr, 1965; Schwartz, 1971*a*, *b*).

(e) Finally, equation (3.11) has been extended to include conditions when interfacial electrical potential differences do not cancel (e.g., the partition coefficients are dependent on concentration) and when the electrical potential profiles within the membrane are nonlinear; however, these treatments, though mathematically sound, are difficult to apply in practice (see Schwartz, 1971*a*, *b*; Jacquez, 1971; Barr, 1965; Jacquez and Schultz, 1974).[4]

3.3. A general expression for a diffusion potential

We now derive an expression for a diffusion potential, building on some of the principles outlined previously, which is generally valid under isobaric, isothermal conditions and does not depend upon assumptions regarding membrane composition, electrical potential profiles, or ionic mobilities. More rigorous derivations may be found elsewhere (Staverman, 1952; Scatchard, 1953; Kirkwood, 1954).

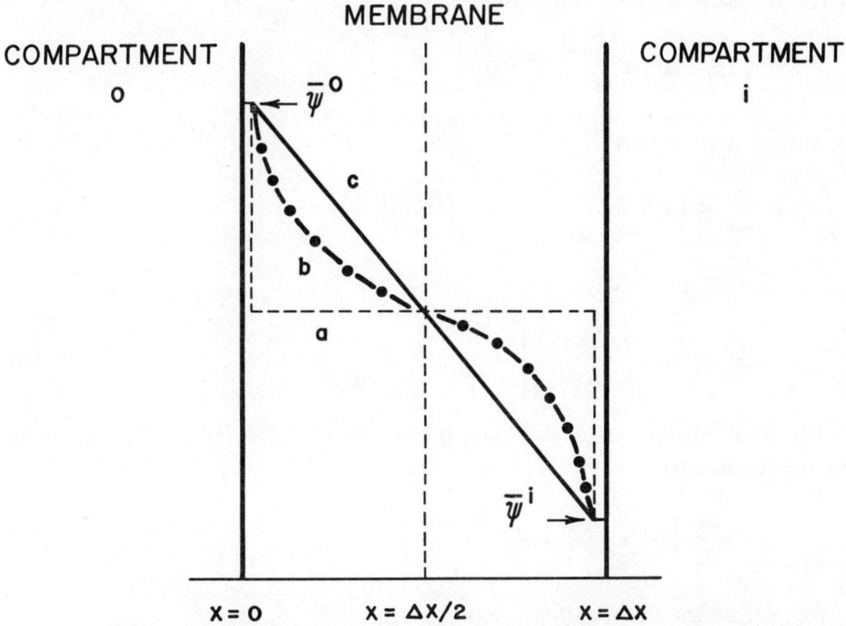

Figure 3.1. Examples of the family of antisymmetric electrical potential profiles that are consistent with the GHK equation (3.11). The constant-field profile is given by line *c*. The other extreme, designated by *a*, is when the electrical potential difference is the result of two interfacial potentials in the same direction; in this case, there is no electrical field acting on the ions during passage through the membrane and if (as shown) the interfacial potential differences are of equal magnitude, J_i is given by equation (2.21) (Mullins, 1961). The curve marked *b* is an intermediate, antisymmetric profile between the extremes *a* and *c*.

Consider the diffusion of *n* charged species across a membrane surrounded by two homogeneous solutions in the same solvent under steady-state conditions, so that the flows and forces are time independent. According to equation (1.36),

$$T(d_i S/dt) = \sum_i^n J_i X_i$$

Now, as discussed by MacInnes (1961), the barrier may be viewed as consisting of an indefinite number of layers, or lamina, so that the differences in the concentrations of *i* across each layer can be made infinitesimally small and the transfer of *i* across each layer can be considered a *reversible* process.

Thus, for *each layer* we can write

$$T(d_iS/dt) = \sum_i^n J_i X_i = 0$$

or from equation (2.4)

$$\sum_i^n J_i \left[RT \left(\frac{d \ln \bar{c}_i}{dx} \right) + z_i \mathcal{F} \left(\frac{d\bar{\psi}}{dx} \right) \right] = 0 \tag{3.15}$$

or

$$\sum_i^n J_i \left[RT \left(\frac{d \ln \bar{c}_i}{dx} \right) \right] + \sum_i^n z_i J_i \mathcal{F} \left(\frac{d\bar{\psi}}{dx} \right) = 0 \tag{3.16}$$

The total current across this layer is given by $I = \Sigma_i^n z_i \mathcal{F} J_i$, so that equation (3.16) reduces to

$$-\left(\frac{d\bar{\psi}}{dx} \right) = \sum_i^n \left(\frac{J_i}{I} \right) RT \left(\frac{d \ln \bar{c}_i}{dx} \right) \tag{3.17}$$

We now define the *transference number* t_i as[5]

$$t_i = (I_i/I) = (z_i \mathcal{F} J_i/I)_{\Delta c_i = 0}$$

Thus, t_i is the fraction of the *total* current across the layer attributable to the flow of i when $\Delta c_i = 0$; clearly, $\Sigma_i\, t_i = 1$. Substituting this expression into equation (3.17), we obtain

$$-\left(\frac{d\bar{\psi}}{dx} \right) = \frac{RT}{\mathcal{F}} \sum_i^n \left(\frac{t_i}{z_i} \right) \left(\frac{d \ln \bar{c}_i}{dx} \right)$$

Integrating this expression, for a single lamina, across the entire thickness of the membrane from solution o to solution i, we obtain

$$\Delta\psi = -\frac{RT}{\mathcal{F}} \int_o^i \left[\sum_i^n \left(\frac{t_i}{z_i} \right) d \ln c_i \right] = -\frac{1}{\mathcal{F}} \int_o^i \sum_i^n \left(\frac{t_i}{z_i} \right) d\mu_i \tag{3.18}$$

It should be emphasized that the integration *includes* the boundary conditions so that equation (3.18) refers to the properties of the external solutions (the initial and final states of the system). A rigorous justification for this procedure is given by Kirkwood (1954); in essence it rests on the fact that the electrical potential difference across any layer is a property of state (i.e., independent of path) so that the overall electrical potential difference ($\Delta\psi$) is simply the integral (summation) of these elemental electrical potential differences. It also follows that although equation (3.18) was derived assuming that

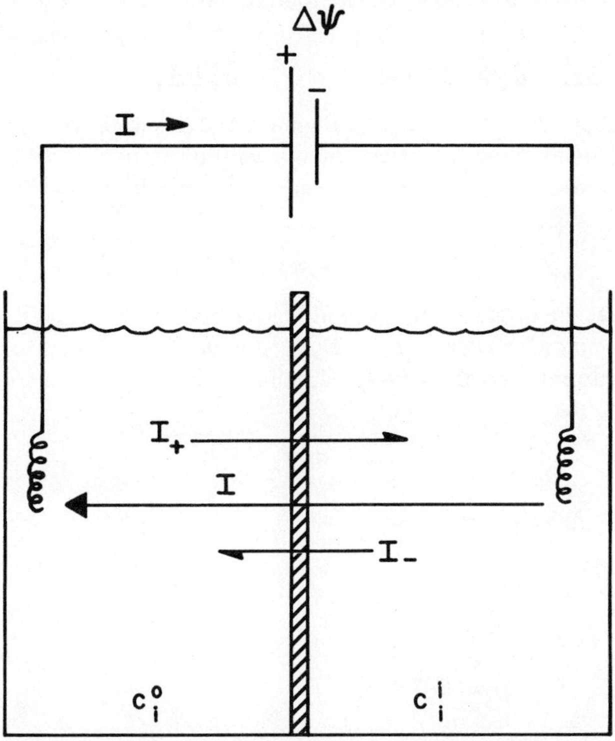

Figure 3.2. System employed for the measurement of transference numbers across a membrane. The total current flowing through the external circuit is the sum of oppositely directed cationic currents and anionic currents. Electroneutrality of the two solutions is maintained by "electrode reactions." For an excellent description of the determination of transference numbers and electrode reactions, see MacInnes (1961, pp. 59–96).

the transfer of matter across any layer can be treated as a reversible process, the final expression for $\Delta\psi$ is generally valid.

However, in order to integrate the right-hand side of equation (3.18), the relation between t_i, the transference number of the *membrane as a whole*, and c_i must be known. In order to illustrate the problem, let us consider the system shown in Figure 3.2, where a membrane separates two identical solutions containing an arbitrary mixture of anions and cations; thus, $c_i^o = c_i^i$ for all i. If we impose an electrical driving force $\Delta\psi$ across the membrane by means of the battery, the total current across the membrane will be made up of a cationic current I_+, directed from the anode to the cathode (from compartment o to compartment i in the figure), and an anionic current I_-, flowing in the opposite direction. Thus, from equation (2.4), we can write for each cation and each anion

$$J_+ = -z_+\bar{c}_+u_+\,\mathcal{F}\Delta\psi \quad \text{and} \quad J_- = -z_-\bar{c}_-u_-\,\mathcal{F}\Delta\psi$$

(recall that $\Delta\psi = \psi^i - \psi^o$, so that in Figure 3.2, $\Delta\psi$ is negative, J_+ is positive, and J_- is negative). Now, we relate the concentrations of i in the membrane to those in the outer solutions by means of the partition coefficients β_i so that

$$J_+ = -z_+\beta_+c_+u_+\,\mathcal{F}\Delta\psi \quad \text{and} \quad J_- = -z_-\beta_-c_-u_-\,\mathcal{F}\Delta\psi$$

It follows that the *total* cationic current is given by $I_+ = -\Sigma_c\, z_+^2\beta_+c_+u_+\mathcal{F}^2\Delta\psi$, and the *total* anionic current is $I_- = -\Sigma_a\, z_-^2\beta_-c_-u_-\mathcal{F}^2\Delta\psi$.

The total current is given by $I = I_+ + I_-$; thus

$$I = -\sum_c z_+^2\beta_+c_+u_+\mathcal{F}^2\Delta\psi - \sum_a z_-^2\beta_-c_-u_-\mathcal{F}^2\Delta\psi$$

or

$$I = -\sum_i z_i^2\beta_ic_iu_i\mathcal{F}^2\Delta\psi$$

Since for any ion i, $I_i = -z_i^2\beta_ic_iu_i\mathcal{F}^2\Delta\psi$, it follows that

$$t_i = \left(\frac{I_i}{I}\right) = \frac{[z_i^2\beta_ic_iu_i]}{\left[\sum_i z_i^2\beta_ic_iu_i\right]} \tag{3.19}$$

Clearly, $\Sigma_i\, t_i = 1$. Thus, in general, t_i is a complex function of concentrations (activities), partition coefficients, and mobilities.

Let us now consider the simple case where the membrane separates two solutions containing a single salt at concentrations c_i^o and c_i^i. Let us further assume (as before) that $z_+ = -z_-$ and that the membrane is "neutral" so that $\beta_+ = \beta_- = \beta_i$ and β_i is independent of concentration. From equation (3.19) it follows that

$$t_+ = \frac{u_+}{u_+ + u_-} \quad \text{and} \quad t_- = \frac{u_-}{u_+ + u_-} \tag{3.20}$$

Thus, if the mobilities are independent of concentration, t_+ and t_- are constants and equation (3.18) can be written as follows:

$$\Delta\psi = -\left(\frac{RT}{z_+\mathcal{F}}\right)t_+\int_o^i d\ln c_+ + \left(\frac{RT}{z_+\mathcal{F}}\right)t_-\int_o^i d\ln c_-$$

Since the preservation of electroneutrality requires that $c_+ = c_- = c_i$ throughout the system, it follows that

$$\Delta\psi = -\left(\frac{RT}{z_+\mathcal{F}}\right)(t_+ - t_-)\int_o^i d\ln c_i \tag{3.21}$$

or, since $t_+ + t_- = 1$,

$$\Delta\psi = \left(\frac{RT}{z_+\mathcal{F}}\right)(1 - 2t_+)\int_o^i d\ln c_i \tag{3.22}$$

Integrating equation (3.21) across the total membrane, we obtain

$$\Delta\psi = -\left(\frac{RT}{z_+\mathcal{F}}\right)(t_+ - t_-)\ln\left(\frac{c_i^i}{c_i^o}\right) \tag{3.23}$$

and substituting the relations between t_i and u_i given by equations (3.20) into equation (3.23) yields

$$\Delta\psi = -\frac{RT}{z_+\mathcal{F}}\left(\frac{u_+ - u_-}{u_+ + u_-}\right)\ln\left(\frac{c_i^i}{c_i^o}\right)$$

which is identical with equation (3.4).

In order to apply equation (3.18) to more complex systems, the relations between t_i and c_i must be determined experimentally. This empirical relation is then substituted into equation (3.18) and the final expression is integrated. This approach has been employed by several groups and excellent agreement has been found between the predicted and observed values of $\Delta\psi$ (Lakshminarayanaiah, 1969; Gunn and Curran, 1971; MacInnes, 1961).

3.4. The Henderson, Planck, and Schlögl equations

As discussed above, equation (3.18) is derived using the Nernst-Planck equation (2.4), which can be solved explicitly under certain restricted conditions. However, in general, the integration of equation (3.18) or equation (2.4) requires some assumption regarding the electrical potential or ionic activity profiles within the membrane. As we have seen, one such assumption is that the electrical field within the membrane is constant (i.e., ψ is a linear function of x), which leads to the GHK equation (3.11).

Henderson (1907, 1908) integrated equation (3.18), assuming constant concentration gradients *within* the membrane; or, more exactly, assuming that the region separating the two surrounding solutions can be viewed as a continuous series of solutions produced by the gradual admixture of the two external solutions. Thus, at any point x within the membrane

$$\bar{c}_i^x = \bar{c}_i^o + (\bar{c}_i^i - \bar{c}_i^o)x/\Delta x$$

and from the definition of t_i, equation (3.19),

$$t_i = \frac{\bar{c}_i^x u_i z_i^2 \Delta x}{(\Delta x - x)\sum_i^n \bar{c}_i^o u_i z_i^2 + x\sum_i^n \bar{c}_i^i u_i z_i^2}$$

If u_i is independent of concentration over the range c_i^o to c_i^i for all ions, equation (3.18) can be integrated to yield the *Henderson diffusion equation*

$$-\Delta\bar{\psi} = \frac{RT}{\mathcal{F}}\left(\frac{\sum_i^n u_i(\bar{c}_i^i - \bar{c}_i^o)}{\sum_i^n u_i(\bar{c}_i^i - \bar{c}_i^o)\, z_i}\right) \ln\left(\frac{\sum_i^n \bar{c}_i^o u_i z_i}{\sum_i^n \bar{c}_i^i u_i z_i}\right) \tag{3.24}$$

If all of the ions are univalent so that $z_+ = 1 = -z_-$, equation (3.24) reduces to

$$-\Delta\bar{\psi} = \frac{RT}{\mathcal{F}}\left[\frac{(U^o - V^o) - (U^i - V^i)}{(U^o + V^o) - (U^i + V^i)}\right] \ln\left(\frac{U^o + V^o}{U^i + V^i}\right) \tag{3.25}$$

where

$$U^o = \sum_c \bar{c}_+^o u_+, \quad V^o = \sum_a \bar{c}_-^o u_-$$
$$U^i = \sum_c \bar{c}_+^i u_+, \quad V^i = \sum_a \bar{c}_-^i u_- \tag{3.26}$$

and Σc is the summation over all cations and Σa is the summation over all anions. For the case of a dissociated univalent-univalent salt, equation (3.25) reduces to equation (3.4).

Planck (1890*a*, *b*) integrated equation (2.4) for the case where $z_+ = -z_- = 1$, assuming that diffusion takes place across a "constrained boundary" separating two well-stirred solutions and that microscopic electroneutrality is closely *approximated* at every point within the barrier; that is, there is no significant space charge density within the membrane or $\Sigma z_i \bar{c}_i \cong 0$. [It should be noted that this assumption differs only slightly from the constant field assumption. Thus, according to the Poisson equation

$$\left(\frac{d^2\psi}{dx^2}\right) = \left(\frac{dE}{dx}\right) = \left(\frac{\mathcal{F}}{\epsilon}\right) \sum_i z_i \bar{c}_i$$

where E is the electric field and ϵ is the permittivity of the barrier. Clearly, when E is constant, $\Sigma_i z_i \bar{c}_i = 0$; this is the condition of *strict* microscopic electroneutrality throughout the membrane.] The result of Planck's integration[6] is the transcendental equation

$$\frac{\xi U^i - U^o}{V^i - \xi V^o} = \left[\frac{\ln(\bar{C}^i/\bar{C}^o) - \ln \xi}{\ln(\bar{C}^i/\bar{C}^o) + \ln \xi}\right]\left(\frac{\xi \bar{C}^i - \bar{C}^o}{\bar{C}^i - \xi \bar{C}^o}\right) \tag{3.27}$$

where U^o, V^o, U^i, V^i are defined as in (3.26); $\bar{C}^o$ and $\bar{C}^i$ are the *total ionic concentrations;* and $\xi = \exp(\mathcal{F}\Delta\bar{\psi}/RT)$.

When all ions are monovalent and $\overline{C}_i^o = \overline{C}_i^i$

$$\frac{\xi U^i - U^o}{V^i - \xi V^o} = 1$$

so that

$$\xi = \frac{V^i + U^o}{V^o + U^i}$$

and

$$\Delta\overline{\psi} = \frac{RT}{\mathcal{F}} \ln \left[\frac{\sum_c \overline{c}_+^o u_+ + \sum_a \overline{c}_-^i u_-}{\sum_c \overline{c}_+^i u_+ + \sum_a \overline{c}_-^o u_-} \right]$$

which is identical with the GHK equation (3.11).

Further, for the case of a single univalent-univalent salt, where $c_i^o \neq c_i^i$, equation (3.27) reduces to equation (3.4).

Equation (3.27) has been extended to the case where all cations have one valence and all anions have another (Johnson, 1904), and a general formulation for arbitrary mixture of ions has been derived by Pleijel (1910).

However, equation (3.27) is rather unwieldy and difficult to apply except under the most simple circumstances; the extensions of this expression are even more cumbersome. Planck suggested a graphical method for solving equation (3.27) for ξ when the compositions of the two bathing solutions and the ionic mobilities are known. Thus, we let

$$\left(\frac{\xi U^i - U^o}{V^i - \xi V^o}\right) = a \quad \text{and} \quad \left[\frac{\ln (\overline{C}^i/\overline{C}^o) - \ln \xi}{\ln (\overline{C}^i/\overline{C}^o) + \ln \xi}\right] \left(\frac{\xi \overline{C}^i - \overline{C}^o}{\overline{C}^i - \xi \overline{C}^o}\right) = b$$

Then a and b are evaluated for different values of ξ and plotted against the chosen values of ξ; the point where a and b intersect defines the value of ξ compatible with equation (3.27).

Finally, Schlögl (1954, 1964) has derived a very general solution of the Nernst–Planck equation for arbitrary mixtures of electrolytes without assumptions regarding the electrical potential or concentration profiles within the membrane. The derivation is very complex and the final expressions are extremely difficult to apply in practice. The reader is referred to the original papers and the brief outline by Helfferich (1962; pp. 386–8) for further details. Suffice it to say that when both solutions contain equal total concentrations of ions having the same valence, the predicted electrical potential profile is linear and the final expressions resemble the constant field equation.

3.5. The Donnan equilibrium and the Teorell-Meyer-Sievers (TMS) model

Up to this point we have limited our considerations to systems where the partition or distribution coefficient between the membrane interface and the adjacent solution β_i is assumed to be independent of concentration so that $\beta_i^o = \beta_i^i$ for all i. However, in general, this assumption is not strictly valid even for so-called neutral membranes and is grossly incorrect for the case of charged membranes that contain a high density of *fixed* (structural) charged (dissociated) groups.[7] In this section we will examine the behavior of charged membranes, where $\beta_i^o \neq \beta_i^i$, and develop a general approach toward the analysis of transmembrane diffusion potentials.

Let us start by considering a system where compartment o contains a solution of a salt that dissociates into two univalent ions at concentrations c_+^o and c_-^o, and compartment i contains a solution of the same ions, c_+^i and c_-^i as well as another charged anion X_-^i with valence z_x. We assume that the intervening membrane is permeable to c_+ and c_- but impermeable to X_-^i. When the system achieves equilibrium, $\tilde{\mu}_+^o = \tilde{\mu}_+^i$ and $\tilde{\mu}_-^o = \tilde{\mu}_-^i$, so that from equation (1.32)

$$\Delta\psi = \psi^i - \psi^o = -(RT/\mathcal{F})\ln(c_+^i/c_+^o) = (RT/\mathcal{F})\ln(c_-^i/c_-^o) \qquad (3.28)$$

We now define $r = (c_+^i/c_+^o) = (c_-^o/c_-^i)$, so that $\Delta\psi = -(RT/\mathcal{F})\ln r$. Now, since both external solutions must be electroneutral,

$$c_+^o = c_-^o = c_i^o \quad \text{and} \quad c_+^i = c_-^i + |z_x|X_-^i \qquad (3.29)$$

Substituting the definition of r and rearranging equations (3.29) readily yields

$$r^2 c_i^o - r|z_x X_-^i| - c_i^o = 0$$

and, solving this quadratic equation, we obtain

$$r = \frac{|z_x X_-^i|}{2c_i^o} + \left[\left(\frac{z_x X_-^i}{2c_i^o}\right)^2 + 1\right]^{1/2} \qquad (3.30)$$

(*Note:* The positive root is chosen because the square-root term must always be larger than the first term and r cannot be negative.)

This equilibrium system is referred to as the *Donnan* (*or Gibbs-Donnan*) *equilibrium* and the value r is often called the *Donnan ratio*.

Now, let us consider a membrane that contains a high density of fixed negative charges that separates two solutions of a dissociated salt c_i, where $c_i^o > c_i^i$ (Figure 3.3). If translocation of ions across the membrane is slow compared to the partition of ions across the interfaces, the interfacial distributions will approach those predicted by the Donnan equilibrium, where β_i is a function of concentration given by equation (3.30) so that $\beta_i^o \neq \beta_i^i$. In

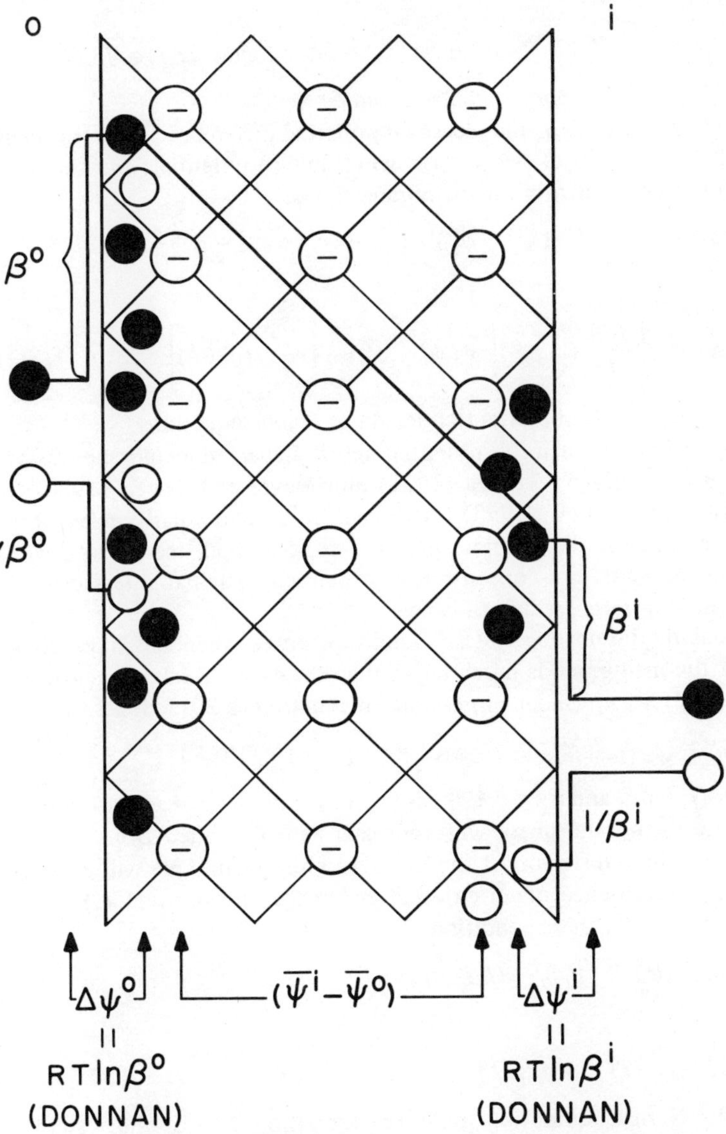

Figure 3.3. The origin of electrical potential differences across a membrane containing fixed anionic charges according to the Teorell–Meyer–Sievers model. The fixed anionic charges are designated by ⊖; mobile cations by ●; and mobile anions by ○.

this instance

$$\bar{c}_+^{\,o} = \beta_i^{\,o} c_+^{\,o};\quad \bar{c}_-^{\,o} = c_-^{\,o}/\beta_i^{\,o};\quad \bar{c}_+^{\,i} = \beta_i^{\,i} c_+^{\,i} \quad\text{and}\quad \bar{c}_-^{\,i} = c_-^{\,i}/\beta_i^{\,i}$$

Clearly, $\bar{c}_+^{\,o} > c_+^{\,o}, \bar{c}_-^{\,o} < c_-^{\,o}, \bar{c}_+^{\,i} > c_+^{\,i}$ and $\bar{c}_-^{\,i} < c_-^{\,i}$.

As shown in Figure 3.3, the electrical potential difference across this membrane will be the sum of two interfacial (Donnan) potentials *and* the diffusion potential arising *within* the membrane. Thus,

$$\Delta\psi = \psi^i - \psi^o = (\bar{\psi}^o - \psi^o) + (\bar{\psi}^i - \bar{\psi}^o) + (\psi^i - \bar{\psi}^i)$$

or

$$\Delta\psi = -\left[\frac{RT}{\mathcal{F}}\,(\ln \beta_i^{\,o})\right] + (\bar{\psi}^i - \bar{\psi}^o) + \left[\frac{RT}{\mathcal{F}}\,(\ln \beta_i^{\,i})\right] \qquad (3.31)$$

where $(\bar{\psi}^i - \bar{\psi}^o)$ is given by the Henderson or Planck equations.

This model for the diffusion potential across a charged membrane was proposed simultaneously by Teorell (1935) and Meyer and Sievers (1936) and later revised by Teorell (1953). Although it is particularly relevant for charged membranes, the general approach applies to all membranes where partition coefficients are concentration dependent, so that interfacial electrical potential differences do not cancel.

A particularly interesting *special* condition arises when the fixed charge density of the membrane is much larger than the external salt concentration, that is, $|z_x \bar{X}_-^{\,i}| \gg c_i$. Under this condition equation (3.30) reduces to

$$\beta_i^{\,o} = (z_x \bar{X}_-/c_i^{\,o}) \gg 1 \quad\text{and}\quad \beta_i^{\,i} = (z_x \bar{X}_-/c_i^{\,i}) \gg 1$$

Since $\bar{c}_-^{\,o} = c_-^{\,o}/\beta_i^{\,o}$ and $\bar{c}_-^{\,i} = c_-^{\,i}/\beta_i^{\,i}$, as $(z_x X_-) \to \infty$, the concentration of the anion in the membrane phase will approach zero (so-called *Donnan exclusion*). In the limit, diffusion of the salt across the membrane will cease (recall that in the absence of an external current, J_+ must equal J_-) so that $(\bar{\psi}^i - \bar{\psi}^o) = 0$. Under these conditions

$$\Delta\psi = (RT/\mathcal{F}) \ln \beta_i^{\,i} - (RT/\mathcal{F}) \ln \beta_i^{\,o}$$

or

$$\Delta\psi = (RT/\mathcal{F}) \ln(c_i^{\,o}/c_i^{\,i})$$

which is the Nernst equilibrium potential (equation 3.7) of the system. In this instance $\Delta\psi$ is the sum of two equilibrium potentials that arise solely at the interfaces of the membrane. This is the origin of the electrical potential differences generated by near perfect ion-exchange membranes and ion-selective materials, for example, pH electrodes (cf. Eisenman, 1967).

4 Water transport

4.1. Osmosis, van't Hoff's law, and Staverman's reflection coefficient

One of the seminal advances in physical chemistry during the nineteenth century was the recognition that solutes in dilute solution behave as a perfect gas; indeed, the extension of the perfect gas law to dilute solutions represents the foundation of the "thermodynamics of solutions." This point can be illustrated by considering two rigid compartments at the same temperature separated by a rigid barrier. Assume that compartment o contains n^o moles of a gas in a volume V^o and that compartment i contains n^i moles of the same gas in a volume V^i. According to the perfect gas law

$$P^o V^o = RTn^o \quad \text{or} \quad P^o = RTc^o$$

where $c^o = n^o/V^o$ and is simply the concentration of the gas in moles per unit volume. Similarly, for compartment i we can write $P^i = RTc^i$. Thus, if the barrier is nonporous,

$$\Delta P = P^o - P^i = RT(c^o - c^i) = RT\Delta c \tag{4.1}$$

This is the pressure difference that will be sensed by pressure transducers inserted into each of the compartments. If the barrier is porous, gas will diffuse from the compartment with the higher concentration (pressure) to that with the lower concentration (pressure) until ΔP and Δc are abolished; thus, ΔP can be considered the "diffusion pressure" of the gas.

Let us now consider the system illustrated in Figure 4.1, where compartment o contains an aqueous solution of solute i at a concentration c_i^o and compartment i contains a solution of the same solute at a *higher* concentration c_i^i; the intervening membrane is assumed to be rigid and strictly impermeable to i but freely permeable to water. To the piston sealing compartment i, we now apply a pressure of sufficient magnitude to prevent any change in volume of either compartment. Since in this system a change in volume can only come about as a result of water flow, the pressure applied must be sufficient to abolish the driving force for water flow so that, from Section 1.5, $\Delta\mu_w = \mu_w{}^i - \mu_w{}^o = 0$, where μ_w is the chemical potential of water. In other words the state properties of the system must be such that water is at equilibrium!

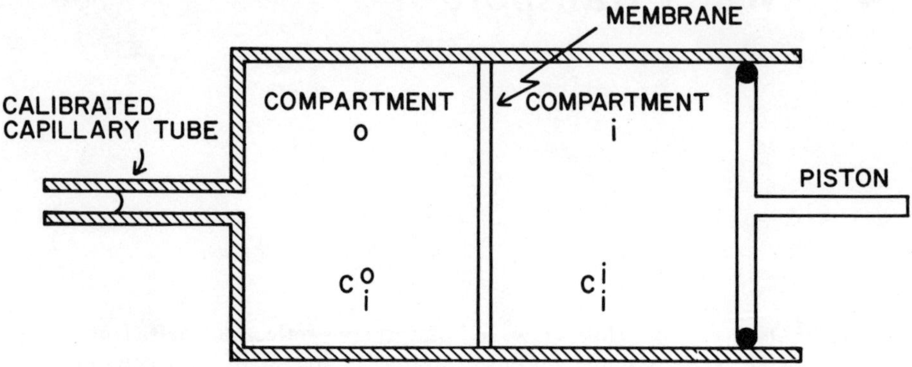

Figure 4.1. System for studying osmotic pressure and osmotic volume flow.

From equation (1.27) we can write

$$\mu_w{}^o = (\mu_w{}^0)^o + \bar{v}_w P^o + RT \ln x_w{}^o \tag{4.2}$$

where $\bar{v}_w$ is the partial molar volume of water (approximately 18 cm^3/mole), P^o is the pressure in compartment *o*, and $x_w{}^o$ is the *mole fraction* of water in compartment *o*;[1] clearly, $x_w{}^o = n_w{}^o/(n_i{}^o + n_w{}^o)$. Similarly, for compartment *i*

$$\mu_w{}^i = (\mu_w{}^0)^i + \bar{v}_w P^i + RT \ln x_w{}^i \tag{4.3}$$

Equating (4.2) and (4.3), substituting the definitions of x_w and solving, assuming that $(\mu_w{}^0)^o = (\mu_w{}^0)^i$, we obtain

$$\bar{v}_w(P^i - P^o) = RT\left[\ln\left(1 + \frac{n_i{}^i}{n_w{}^i}\right) - \ln\left(1 + \frac{n_i{}^o}{n_w{}^o}\right)\right] \tag{4.4}$$

Since, in dilute aqueous solution, $n_w \gg n_i$, and since $\ln(1 + x)$ approaches x as x approaches zero, equation (4.4) simplifies to

$$\bar{v}_w(P^i - P^o) = RT\left(\frac{n_i{}^i}{n_w{}^i} - \frac{n_i{}^o}{n_w{}^o}\right) \tag{4.5}$$

Since $\bar{v}_w n_w{}^i \cong V^i$ and $\bar{v}_w n_w{}^o \cong V^o$, where V is the volume of the compartment, equation (4.5) finally reduces to

$$\pi = (P^i - P^o) = RT(c_i{}^i - c_i{}^o) = RT\Delta c_i \tag{4.6}$$

which is the van't Hoff equation for the osmotic pressure (π) across a barrier separating two ideal, dilute solutions *when the barrier is strictly impermeable to the solute.* Equation (4.6) is identical with equation (4.1) derived for perfect gases. In this case pressure is applied to the more concentrated solution (of solute) which is obviously the side with the lower concentration of water; thus, in strict analogy with the diffusion of gases, π can be referred to as the "diffusion pressure of water."

It should be emphasized that the system we have just discussed is at equilibrium and π is the pressure that must be applied to the *concentrated* solution to offset, precisely, the difference in water activities across the barrier; in this sense equation (4.6) is analogous to the Nernst equation (2.31; 3.7), where the electrical potential difference across the barrier precisely offsets the difference in activities of a permeant ion. In both instances the equilibrium condition is achieved by a balancing of different but equivalent driving forces.

If the piston illustrated in Figure 4.1 is released so that $(P^i - P^o) = 0$ (i.e., both chambers are now open to the atmosphere), water will flow from o to i, and the initial rate of water flow (i.e., before there are any changes in the concentrations) is given by

$$\bar{v}_w J_w = J_v = L_p \pi \tag{4.7}$$

where L_p is the *hydraulic conductivity* of the membrane; for a barrier permeated by right cylindrical pores whose diameter is much larger than that of water, L_p is given by the Poiseuille equation. Equation (4.7) is an example of the linear relation between flows and forces discussed in general terms in Chapter 1.

More than a half-century elapsed between the formulation of van't Hoff's law and the next major advance in the theory of osmotic pressure and osmotic water flow; namely, the generalization of equation (4.6) by Staverman to include membranes that are permeable to the solute (Staverman, 1951).

Staverman reasoned as follows: If compartments o and i in Figure 4.1 are filled with the same aqueous solution with a concentration c_i, then the application of a pressure to the piston will force volume from i to o, and the "ultrafiltration permeability" of i can be calculated from the change in V^o and c_i^o. For example, if the membrane is impermeable to i, then only pure water will be forced into compartment o and after a given period, the initial c_i^o will decline to a value given by $(c_i^o)' = (c_i^o V^o)/(V^o + \Delta V^o)$, where ΔV^o is the increment in volume. Conversely, if the membrane is *equally permeable* to solute and water, the solution in compartment i will simply be transferred, unaltered, to compartment o and c_i^o will not change. Obviously, if the membrane is permeable to i but less so than to water, the change in c_i^o will fall between these extremes. The ability of the barrier to discriminate between the solute and water can be described by a *reflection coefficient* σ defined as

$$\sigma_i = 1 - (c_i^{\text{filtrate}}/c_i^{\text{filtrand}}) \tag{4.8}$$

where for small volume flows, c_i^{filtrand} is simply c_i^i and c_i^{filtrate} is the calculated composition of the solution that was forced into compartment o.

Clearly, if the membrane is impermeable to the solute $\sigma_1 = 1$ and, if the membrane is equally permeable to solute and water, $\sigma_i = 0$; for all intermediate cases $0 < \sigma_i < 1$. Clearly, σ_i is a function of both the solute i and the membrane. The term "reflection coefficient" was chosen because it indicates the extent to which the solute particles are "reflected" by the membrane in comparison with the solvent.

Staverman then showed that if $c_i^i > c_i^o$ the *initial* pressure that must be applied to the piston to prevent volume flow is given by

$$\pi_{\text{eff}} = \sigma_i \pi = \sigma_i RT\Delta c_i \tag{4.9}$$

and the initial rate of volume flow from o to i when $\Delta P = 0$ is

$$J_v = \sigma_i L_p RT\Delta c_i \tag{4.10}$$

where π_{eff} is the *effective osmotic pressure* across a nonideal membrane.

Equations (4.9) and (4.10) are completely generalized descriptions of osmotic pressure and osmotic volume flow.

As we will see, σ_i is a "cross coefficient" that is the measure of the coupling (interaction) between the flows of solute and solvent across a membrane; and σ_i together with the two "straight coefficients" P_i and L_p are sufficient to completely describe the flows of solute and solvent across a given membrane.

4.2. Interactions between solute and solvent flows: The Kedem–Katchalsky equations

As implied in Section 4.1, when a membrane that is permeable to both solute and solvent separates two solutions having different concentrations, there will be oppositely directed flows of solute and solvent, and as discussed in Section 1.6, in general, these two flows can interact. In 1958 Kedem and Katchalsky (K and K) analyzed such systems using the formalism of irreversible thermodynamics; in one fell swoop they interjected this relatively young discipline into the biological literature, and their contribution has had a profound influence on the conceptualization and investigation of membrane transport processes.

As discussed in Section 1.6, in a system characterized by two flows and two forces, the "dissipation function" is given by

$$\Phi = T\, d_iS/dt = J_i X_i + J_w X_w \tag{4.11}$$

where, in the system we will consider, J_i is the flow of an uncharged solute and J_w is the flow of water. Therefore, X_i is simply $\Delta\mu_i$ and X_w is $\Delta\mu_w$. We will define the flows as positive when directed from compartment o to compartment i and $\Delta\mu_i = \mu_i^o - \mu_i^i$ and $\Delta\mu_w = \mu_w^o - \mu_w^i$.

From equations (4.2) to (4.6) it is a simple matter to show that for a dilute solution where $n_i/n_w \to 0$ and $n_w\bar{v}_w \to V$

$$\Delta\mu_w = \bar{v}_w(P^o - P^i) - \bar{v}_w RT(c_i^o - c_i^i) = \bar{v}_w(\Delta P - \pi) \tag{4.12}$$

For the case of the solute i, from equation (1.27),

$$\Delta\mu_i = \bar{v}_i\,\Delta P + RT\Delta \ln c_i \qquad (4.13)$$

where $\Delta \ln c_i = \ln c_i^o - \ln c_i^i = \ln(c_i^o/c_i^i)$.

In order to linearize equation (4.13) (i.e., eliminate the logarithmic term), K and K introduced the notion of a *mean concentration* $\dot{c}_i$, defined by

$$\dot{c}_i = \left(\frac{c_i^o - c_i^i}{\ln(c_i^o/c_i^i)}\right) = \left(\frac{\Delta c_i}{\ln(c_i^o/c_i^i)}\right) \qquad (4.14)$$

Using this definition, equation (4.13) simplifies to

$$\Delta\mu_i = \bar{v}_i\,\Delta P + RT\Delta c_i/\dot{c}_i = \bar{v}_i\,\Delta P + \pi/\dot{c}_i \qquad (4.15)$$

Now, what is the meaning of $\dot{c}_i$? When is this important simplification valid? Using the series expansion of $\ln(c_i^o/c_i^i)$, it can be readily shown that when the concentration difference is small so that $\Delta c_i \to 0$ and $(c_i^o/c_i^i) \to 1$, then $\dot{c}_i \to (c_i^o + c_i^i)/2$; in other words, when the concentration difference is small, $\dot{c}_i$ is the *arithmetic mean* of the concentrations in compartments o and i.

Substituting equations (4.12) and (4.15) into equation (4.11), we obtain

$$T\,d_iS/dt = J_i(\bar{v}_i\Delta P + \pi/\dot{c}_i) + J_w\bar{v}_w(\Delta P - \pi) \qquad (4.16)$$

which upon rearrangement yields

$$T\,d_iS/dt = (J_i\bar{v}_i + J_w\bar{v}_w)\,\Delta P + [(J_i/\dot{c}_i) - J_w\bar{v}_w]\,\pi \qquad (4.17)$$

Now, $(J_i\bar{v}_i + J_w\bar{v}_w)$ is simply the flow of volume J_v. Further, since in dilute solution $\dot{c}_w\bar{v}_w \cong 1$, equation (4.17) can be written as follows

$$T\,d_iS/dt = J_v\,\Delta P + J_D\pi \qquad (4.18)$$

where $J_D = [(J_i/\dot{c}_i) - (J_w/\dot{c}_w)]$ and is simply the difference between the velocities of the oppositely directed flows of solute and solvent. J_D is thus a measure of the velocity of *exchange* of matter across the barrier.

Thus, equation (4.11) has been transformed into a new expression of flows and forces where ΔP is the conjugate driving force for the flow of volume J_v, and π or $RT\Delta c_i$ is the driving force for J_D.

We can now write the linear phenomenologic equations that describe this system as follows:

$$J_v = L_p\,\Delta P + L_{pD}RT\Delta c_i$$

and (4.19)

$$J_D = L_{Dp}\,\Delta P + L_D RT\Delta c_i$$

where, according to the Onsager relation, $L_{pD} = L_{Dp}$.

Although equations (4.19) seem abstract, they, in fact, more faithfully describe the actual experimental system than does equation (4.16). Thus, when $\Delta c_i = 0$ and a pressure is applied to compartment i, we *measure* the rate of increase in the *volume* of compartment o or J_v (*not* J_i or J_w); the ratio $(J_v/\Delta P)_{\pi=0}$ is the hydraulic conductivity L_p. We can also measure the *change* in composition of compartment o due to the fact that the applied pressure will, in general, force a fluid across the membrane that does not have the same composition as the solution in compartment i. This differential effect of an applied pressure on the flows of solute and solvent is commonly referred to as *ultrafiltration;* J_D is a measure of this differential flow and the ratio $(J_D/\Delta P)_{\pi=0} = L_{Dp}$ is a measure of the ultrafiltration properties of the membrane.

Conversely, if $\Delta P = 0$ and $c_i^o \neq c_i^i$, there will be a displacement of volume because of the flow of water (osmosis) from the compartment with a lower concentration of i to the other *and* an oppositely directed flow of volume caused by the diffusion of i down its concentration gradient. What is actually measured is the *total displacement of volume*, and the ratio $(J_v/\pi)_{\Delta P=0} = L_{pD}$ is a measure of the relative contributions of solvent flow and solute flow to the total displacement of volume when $\Delta P = 0$.

If the system is not too far displaced from equilibrium (i.e., if the flows and forces are small) $L_{pD} = L_{Dp}$, so that

$$(J_v/\pi)_{\Delta P=0} = (J_D/\Delta P)_{\pi=0} \tag{4.20}$$

Thus, osmotic *volume* flow in the absence of a hydrostatic pressure is equal to the "ultrafiltration effect" in the absence of a concentration difference. This is precisely the reasoning behind the approach employed by Staverman to determine the reflection coefficient σ_i. Further, from equation (4.19), we see that when $\Delta c_i \neq 0$, then $J_v = 0$ when

$$(\Delta P)_{J_v=0} = -(L_{pD}/L_p)\,\pi \tag{4.21}$$

Comparison of equation (4.21) with equation (4.9) reveals that $\sigma_i = -(L_{pD}/L_p)$.

Finally, when $\Delta P = 0$ and $\Delta c_i \neq 0$, there will be an exchange flow of solvent and solute given by J_D; the membrane property that describes this exchange is $(J_D/\pi)_{\Delta P=0} = L_D$.

We now examine the relations that prevail if the membrane is ideally semipermeable, that is, permeable to solvent (water) but not solute (i). Clearly, from equation (4.9), $\sigma_i = 1$ so that $-L_{pD} = L_p$. In addition, under these conditions $J_D = -(J_w/\dot{c}_w) = -(J_w\bar{v}_w) = -J_v$. Since $(J_D/\pi)_{\Delta P=0} = L_D$, it follows that $L_D = L_p$. Thus, for a semipermeable membrane, $-L_{pD} = L_p = L_D$.

What if the membrane cannot discriminate between solute and solvent? From equations (4.19) we see that when $\Delta c_i = 0$

$$(J_D/J_v)_{\pi=0} = L_{pD}/L_p = -\sigma_i \tag{4.22}$$

Substituting the definitions of J_D and J_v, we obtain

$$\left(\frac{(J_i/\dot{c}_i) - (J_w/\dot{c}_w)}{J_i\bar{v}_i + J_w\bar{v}_w}\right)_{\pi=0} = -\sigma_i$$

and, since in dilute solutions $J_w\bar{v}_w \gg J_i\bar{v}_i$, equation (4.22) reduces to

$$\frac{(J_i/\dot{c}_i)}{(J_w/\dot{c}_w)} = 1 - \sigma_i \tag{4.23}$$

Now, $J_i/\dot{c}_i = v_i$ and $J_w/\dot{c}_w = v_w$, where v_i and v_w are the velocities of solute and solvent molecules, respectively, (see Section 2.1). Thus,

$$(v_i/v_w)_{\pi=0} = 1 - \sigma_i$$

Clearly, if the barrier cannot discriminate between solute and solvent, there can be no "ultrafiltration" in response to an applied pressure so that $v_i = v_w$ and $\sigma = 0$. (Needless to say, if the barrier is ideally semipermeable, $v_i = 0$ and $\sigma = 1$.)

The equations developed here lead directly to two expressions that completely describe the interactions between solute and solvent flow across a *homogeneous* membrane that separates two solutions. The introduction of the definition of σ_i into equation (4.19) yields

$$J_v = L_p(\Delta P - \sigma_i RT\Delta c_i) \tag{4.24}$$

so that J_v is directly proportional to the difference between an applied hydrostatic pressure and an oppositely directed *effective osmotic pressure.* In general, when there are n solutes

$$J_v = L_p\left(\Delta P - \sum_i^n \sigma_i RT\Delta c_i\right)$$

Further, solving equation (4.24) for ΔP and substituting this expression into the equation for J_D (4.19) yields

$$J_i = (1 - \sigma_i)\,\dot{c}_i J_v + \omega_i RT\Delta c_i \tag{4.25}$$

where $\omega_i = \dot{c}_i(L_pL_D - L_{pD}{}^2)/L_p$. Since L_pL_D is always greater than $L_{pD}{}^2$ (Section 1.6), ω_i is always positive. Further, when $J_v = 0$, $(J_i)_{J_v=0} = \omega_i RT\Delta c_i$. Thus, ω_i is a measure of solute permeability and is related to the more familiar permeability coefficient by $P_i = \omega_i RT$.

When $\Delta c_i = 0$ then $J_i = (1 - \sigma_i)\,c_i J_v$ or from equation (4.24) $J_i = (1 - \sigma_i) \cdot c_i L_p\,\Delta P$. Thus, if $\sigma_i \neq 1$, there will be a flow of solute that is directly proportional to the flow of volume or, alternatively, the applied hydrostatic pressure (ΔP). This component of solute flow is frequently referred to as "solvent drag," implying that it arises from an entrainment of solute particles in a flowing stream of solvent. However, as we will see, this interpretation implies that solute and solvent share a common pathway through the membrane (e.g., a pore) and need not be correct.

When $\Delta P = 0$, $-J_v = L_p \sigma_i RT \Delta c_i$ and, substituting this expression into equation (4.25), we obtain

$$J_i = [-(1 - \sigma_i)\,\sigma_i L_p \dot{c}_i + \omega_i] RT \Delta c_i \tag{4.26}$$

4.3. Physical interpretations of L_p, ω, and σ

It should be clear from equation (4.24) that L_p is a straight coefficient that relates the rate of volume flow (which in a dilute aqueous solution is essentially equal to the rate of water flow) to the conjugate driving force, namely, the effective pressure difference across the membrane. Similarly, from equation (4.25), we see that ω_i is also a straight coefficient that relates the diffusional flow of an uncharged solute to its conjugate driving force Δc_i. In contrast σ_i is a cross coefficient that provides a measure of the interaction between the flows of solute and solvent and nonconjugate forces. It should be emphasized that these coefficients are strictly phenomenologic and are independent of underlying mechanism; that is, they are not based on any particular model of the membrane or the flow processes. Although this assures general validity, it is intellectually unsatisfying and, over the years, several attempts have been made to provide these coefficients with intuitively satisfactory physical interpretations. The most successful approach is based on the assumption that the generalized phenomenological resistances or conductances can be translated into mechanical frictional coefficients. Thus, it is argued that when a particle achieves a steady-state velocity (i.e., it is not accelerating) the total force acting on the particle is zero and the thermodynamic driving force X_i is precisely balanced by opposing frictional forces. Thus, we can write $X_i = -\Sigma_j F_{ij}$, where the summation includes all possible frictional forces encountered by the moving particle. The frictional forces are given by $F_{ij} = -f_{ij}(v_i - v_j)$, where f_{ij} is a frictional coefficient and $(v_i - v_j)$ is the difference between the velocities of the solute i and some component j in the system with which it interacts. For the case of the flows of an uncharged solute and water across a membrane, the membrane is generally considered to be the stationary reference for the flows so that $v_m = 0$ and the four frictional coefficients that describe the system are f_{iw}, f_{wi}, f_{im}, and f_{wm}, where f_{iw} is the frictional coefficient between solute and water, and so forth. Using this approach, Kedem and Katchalsky (1961) have derived the following relations:

$$\text{(a)} \quad L_p = \varphi_w \bar{v}_w / f_{wm} \Delta x \tag{4.27}$$

where φ_w is the volume fraction of water in a membrane whose thickness is Δx.

$$\text{(b)} \quad \omega_i = \beta_i / (f_{iw} + f_{im}) \Delta x \tag{4.28}$$

where β_i is the partition coefficient described earlier (Sections 1.5 and 2.2) and

$$\text{(c)} \quad \sigma_i = 1 - \frac{\omega_i \bar{v}_i}{L_p} - \frac{\omega_i f_{iw} \Delta x}{\varphi_w} \tag{4.29}$$

The expressions for L_p and ω_i have relatively straightforward interpretations and are intuitively satisfying (however, see Section 4.5). The expression for σ_i, however, provides important insight into the meaning of this cross coefficient and warrants some discussion.

Consider a lipid, porous membrane that separates two aqueous solutions of solute i. Assume that water can cross the membrane only through the pores and that i, a large lipophilic molecule, is excluded from the pores but can diffuse across the membrane through the lipid matrix. Under these conditions $f_{iw} = 0$ and equation (4.29) reduces to

$$\sigma_i = 1 - (\omega_i \bar{v}_i / L_p) \tag{4.30}$$

When $\Delta P = 0$, then $J_v = -L_p \sigma_i RT \Delta c_i$ [equation (4.24)] so that

$$-J_v = L_p RT \, \Delta c_i - \bar{v}_i \omega_i RT \Delta c_i$$

or (4.31)

$$-J_v = L_p RT \, \Delta c_i - \bar{v}_i J_i$$

Now, the first term on the right of equations (4.31) is simply the volume displacement caused by osmotic water flow and the second term is simply the volume of solute that diffuses across the membrane driven by Δc_i. Obviously, if ω_i or $P_i = 0$, equation (4.31) reduces to the van't Hoff expression for an ideally semipermeable membrane. When $\omega_i > 0$, the deviation from ideal behavior is solely because of the fact that the ideal osmotic volume flow is reduced by the volume flow of the permeant solute in the opposite direction.

Consider the situation when $\Delta c_i = 0$ and a hydrostatic pressure ΔP is applied across the membrane. Under these conditions

$$J_v = L_p \Delta P \quad \text{and} \quad J_i = (1 - \sigma_i)\, c_i J_v$$

Solving equation (4.30) for L_p and substituting, we obtain

$$J_i = \omega_i c_i \bar{v}_i \Delta P \tag{4.32}$$

Now, recall that $\Delta \mu_i = \bar{v}_i \Delta P + RT \Delta \ln c_i$ [equation (4.13)] so that when $\Delta c_i = 0$, $\Delta \mu_i = \bar{v}_i \Delta P$. Thus, equation (4.32) simply states that the flow of i is the product of a mobility term, a concentration term and a driving force (see Section 2.1), where the latter is derived solely from the applied pressure.

Clearly, under these conditions J_i is directly proportional to ΔP and J_v, criteria that are often applied to infer the presence of "solvent drag." However, obviously, there is *no* entrainment of solute in a solvent stream so that this inference is entirely unjustified.

Referring once more to equation (4.29), we see that when the permeant solute and water traverse the membrane through a common path, there are *two factors* that contribute to the decrease in volume flow from that predicted by the van't Hoff equation for an ideal semipermeable membrane. The first is the oppositely directed volume flow of solute and the second arises from the *frictional interaction* between the oppositely directed flows of solvent and solute. Since σ_i, ω_i, L_p, and $\bar{v}_i$ can be determined experimentally, equation (4.29) provides a criterion for determining whether there are interactions between solute and solvent flows across the barrier. Thus, if

$$\sigma_i < 1 - (\omega_i \bar{v}_i / L_p) \qquad (4.33)$$

then $f_{iw} > 0$ and a common path may be inferred.

Finally, according to equation (4.29), σ_i decreases with increasing ω_i and may assume negative values when ω_i is sufficiently large. This situation is referred to as a *negative* or *anomalous* osmosis, inasmuch as J_v will be directed from the compartment with the higher solute concentration to the compartment with the lower solute concentration and if $c_i^i > c_i^o$ (Figure 4.1), a negative pressure would have to be applied to the piston to prevent volume flow; this apparent anomaly, which has been observed, is accounted for by the fact that volume flow of the diffusing solute as well as that resulting from the frictional interaction between solute and solvent overweighs the normally directed osmotic flow (Grim and Sollner, 1957, 1960; Talen and Staverman, 1965; Lakshminarayanaiah, 1969).

In summary, this frictional analysis permits us to express phenomenological coefficients in terms that provide some intuitive feeling for their bases. The principal shortcoming, however, is that only f_{iw} and f_{wi} can be evaluated independently and, then, only under special conditions. Because f_{iw} is the frictional coefficient between a mole of solute and an infinite amount of water, it can be estimated from the diffusion coefficient of i in free solution using the Einstein relation:

$$D_i^o = RTu_i = RT/f_{iw}^o$$

where the superscript o denotes that the process is taking place in free solution. If we assume that the frictional interaction between solute and water in the membrane is equal to that in free solution (e.g., if i diffuses through pores that are large compared to the size of the solute and solvent molecules), then $f_{iw}^o = f_{iw}$. Further, as discussed by Kedem and Katchalsky (1961), f_{wi} (the frictional coefficient between 1 mole of water and an infinite amount of solute) is equal to $\bar{c}_i f_{iw}/\bar{c}_w$. Thus, under the unlikely condition that water-solute and solute-water interactions in the membrane do not differ markedly from those in free solution, f_{iw} and f_{wi} can be approximated.

On the other hand, there are no generally acceptable approaches that permit independent estimates of f_{im} (the frictional coefficient between 1 mole of i and a membrane of infinite thickness) or f_{wm} from first principles; consequently, although the frictional analysis seems intuitively reasonable, it remains phenomenologic.

The Renkin–Durbin model

Renkin and Durbin have suggested a somewhat different approach for the physical interpretation of σ that is also based on hydrodynamic reasoning but is couched in terms that are physically more meaningful than those of the frictional approach. Renkin (1954) argued that the ultrafiltration of an uncharged particle through a membrane permeated by right cylindrical pores could be described by the combined expressions of Ferry (1936) and Faxen (1923) and derived the expression

$$(A_{if}/A_p) = 2[1 - (r_i/r_p)]^2 - [1 - (r_i/r_p)]^4 \cdot [1 - 2.104(r_i/r_p) + 2.09(r_i/r_p)^3 - 0.95\,(r_i/r_p)^5]$$

where (A_{if}/A_p) is the effective area for ultrafiltration divided by the total pore area, and r_i and r_p are the radii of i and of the pore, respectively. As discussed in Section 2.7, A_{if} includes the restrictions upon the entry of i into the pore as well as the hindrance to the flow of i through the pore.

A similar expression can be written for the filtration of water (A_{wf}/A_p) by simply substituting r_w for r_i.

Durbin (1960) later argued that

$$\sigma_i = 1 - (A_{if}/A_{wf})$$

and, therefore, can be defined entirely in terms of r_i, r_w, and r_p. Studies using artificial membranes with pore radii estimated as ranging between 15 Å and 82 Å are reasonably consistent with this analysis (Durbin, 1960). However, it should be emphasized that the smallest pore radius is approximately 10 times that of water (1.5 Å), and it is not clear whether these hydrodynamic models apply when the pore radius approaches the dimensions of the permeating molecules.

4.4. Electrokinetic phenomena

In Section 4.3 we considered the interactions between the flow of solvent (water) and that of an uncharged solute i across a *homogeneous* membrane. We now briefly consider the case when i is a completely dissociable salt; the approach will follow that of Kedem and Katchalsky (1963a), but for the sake of simplicity we will only consider a salt that dissociates into a monovalent cation (+) and a monovalent anion (−) [see also House (1974) for a similar approach].

The experimental system we will refer to is illustrated in Figure 4.2. The

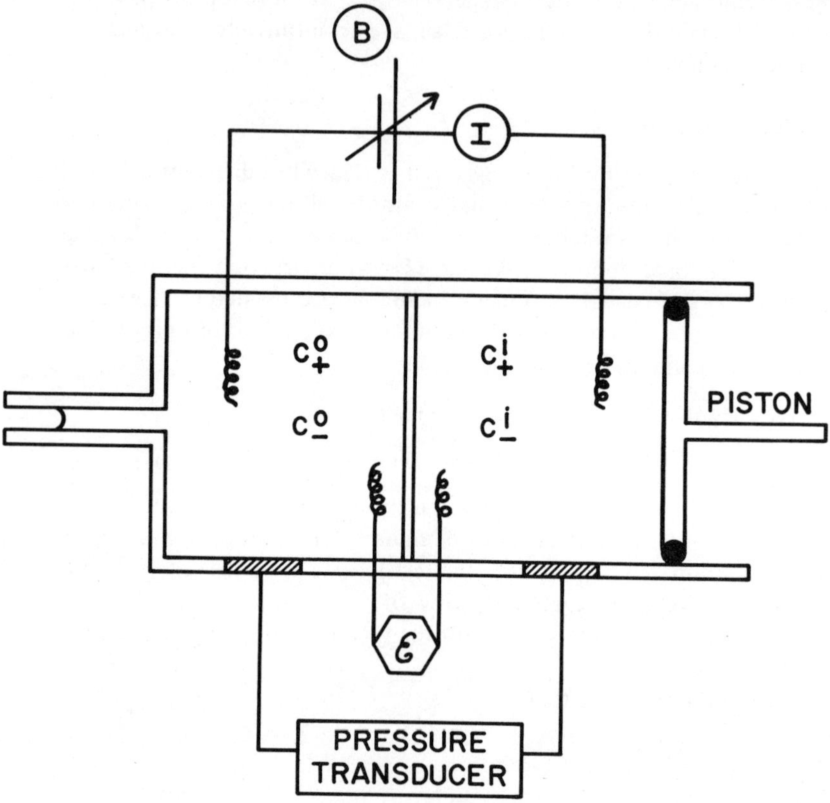

Figure 4.2. System for studying electrokinetic phenomena. The electrodes are reversible to the anion so that the voltmeter $\mathcal{E}$ records the electromotive force of the anion given by equation (4.38). Current I can be passed across the membrane from the external electromotive force (battery) B.

apparatus permits measurement of J_v and ΔP. In addition, electrodes inserted into each compartment connected to a high resistance electrometer (voltmeter) permit measurement of the electrical potential difference across the membrane $\Delta\psi$ under conditions of "essentially zero" current flow. And, identical electrodes connected to a variable electromotive force (battery) are used to pass current I across the membrane. Thus, we can measure $\Delta\psi$ when $I = 0$ as well as the "external current" needed to "clamp" $\Delta\psi$ to zero. For the sake of convenience and in order to adhere to other treatments of this subject in the literature, in this section we define $\Delta\psi = \psi^o - \psi^i$ and continue to define $\Delta c_i = c_i^{\,o} - c_i^{\,i}$, $\Delta P = P^o - P^i$, and $\Delta\mu_i = \mu_i^{\,o} - \mu_i^{\,i}$.

The dissipation function for this tertiary system (i.e., three species, three flows, and three forces) is

$$T\,d_iS/dt = J_+\Delta\tilde{\mu}_+ + J_-\Delta\tilde{\mu}_- + J_w\Delta\mu_w \tag{4.34}$$

where $\Delta\tilde{\mu}_+ = \bar{v}_+\Delta P + RT\Delta \ln c_+ + \mathcal{F}\Delta\psi$ or, using the linearizing approximation discussed in Section 4.3 [equation (4.14)]

$$\Delta\tilde{\mu}_+ = \bar{v}_+\Delta P + (RT\Delta c_+/\dot{c}_+) + \mathcal{F}\Delta\psi \tag{4.35}$$

Similarly,

$$\Delta\tilde{\mu}_- = \bar{v}_-\Delta P + (RT\Delta c_-/\dot{c}_-) - \mathcal{F}\Delta\psi \tag{4.36}$$

Since bulk electroneutrality must be maintained in both compartments, it follows that

$$c_+^{\,o} = c_-^{\,o} = c_i^{\,o} \quad \text{and} \quad c_+^{\,i} = c_-^{\,i} = c_i^{\,i}$$

so that the difference in *chemical potential* of the 1 : 1 salt is

$$\Delta\mu_i = \Delta\tilde{\mu}_+ + \Delta\tilde{\mu}_- = \bar{v}_i\Delta P + 2(RT\Delta c_i/\dot{c}_i) \tag{4.37}$$

where $\bar{v}_i = \bar{v}_+ + \bar{v}_-$. Equation (4.37) simply states that the chemical potential of a dissociable salt is related to the concentration of individual particles resulting from dissociation (i.e., a colligative property of solutions).

Now, the expressions for $\Delta\tilde{\mu}_+$ and $\Delta\tilde{\mu}_-$ given by equations (4.35) and (4.36) could be substituted into equation (4.34) to give an explicit statement of the dissipation function. However, it is more convenient to define the electromotive force of the system, $\mathcal{E}$, which can be measured by inserting identical electrodes *reversible* to one of the ions in each solution and measuring the electrical potential difference between these electrodes potentiometrically (i.e., using a very high resistance voltmeter (electrometer) that draws very little current). The observed electrical potential difference is the *potential* driving force (energy) for the flow of that ion through the external circuit and, if the electrodes are reversible for the anion, is given by (Katchalsky and Curran, 1965, pp. 149-50)

$$\mathcal{E} = -(\Delta\tilde{\mu}_-/\mathcal{F}) = -(RT/\mathcal{F})\,\Delta \ln c_- + \Delta\psi \tag{4.38}$$

in units of volts. Clearly, in any system under steady-state conditions, $\mathcal{E}$ bears a constant relation to $\Delta\psi$.

Now, if the electrodes used to measure the electrical potential difference and to pass current across the membrane are strictly reversible to the anion, the cation is neither withdrawn from nor released into the bathing solutions by the electrode reactions,[2] so that J_+ *across the membrane* can be equated with the *net* displacement of the undissociated salt i, and $J_+ = J_i$.

Further, the current that must be passed across the membrane in order to clamp $\mathcal{E}$ to zero is the difference between the *transmembrane* flows of cation and anion which is equal to the flow of anion (electrons) from one compartment to the other through the external circuit.[3] Thus, the current *through* the membrane is given by

$$I = \mathcal{F}(J_+ - J_-) \tag{4.39}$$

Substituting equations (4.37), (4.38), and (4.39) into equation (4.34) yields a new expression for the dissipation function given by

$$T\,d_iS/dt = J_w\,\Delta\mu_w + J_i\,\Delta\mu_i + I\mathcal{E} \tag{4.40}$$

It follows that we may describe this system by means of the following three linear phenomenological equations:

$$\begin{aligned} J_w &= L_{ww}\,\Delta\mu_w + L_{wi}\,\Delta\mu_i + L_{wI}\mathcal{E} \\ J_i &= L_{iw}\,\Delta\mu_w + L_{ii}\,\Delta\mu_i + L_{iI}\mathcal{E} \\ I &= L_{Iw}\,\Delta\mu_w + L_{Ii}\,\Delta\mu_i + L_{II}\mathcal{E} \end{aligned} \tag{4.41}$$

These equations contain nine phenomenological coefficients; but if Onsager's reciprocal relations are obeyed, $L_{wi} = L_{iw}$, $L_{wI} = L_{Iw}$, and $L_{iI} = L_{Ii}$ so that the number of independent coefficients is reduced to six. Thus, in general, the total water flow is comprised of a hydraulic component, an osmotic component, and a component coupled to the flow of current (or, alternatively, driven by the electromotive force), which is referred to as *electro-osmosis.* Similarly, the total flow of salt is made up of a fraction resulting from ultrafiltration, a fraction resulting from diffusion, and a fraction driven by $\mathcal{E}$ that is referred to as *electrophoresis.* Finally, the total current is made up of a *streaming current* coupled to the flow of water, a diffusion current driven by the concentration difference of the salt and the (electrical) current actually driven by the electromotive force.

Although equations (4.41) completely describe this system, as discussed in Section 4.2, it is frequently more convenient to deal with volume flow, which is actually measured, rather than water flow. Recall that $\Delta\mu_w = \bar{v}_w \cdot (\Delta P - RT\Delta c_i)$ and $\Delta\mu_i = \bar{v}_i\,\Delta P + RT\Delta c_i/\dot{c}_i$, so that equation (4.41) can be written

$$\frac{T\,d_iS}{dt} = J_v(\Delta P - RT\Delta c_i) + \frac{J_i(1 + \dot{c}_i\bar{v}_i)\,RT\,\Delta c_i}{\dot{c}_i} + I\mathcal{E} \tag{4.42}$$

and since in dilute solution $\dot{c}_i\bar{v}_i \ll 1$, equation (4.42) reduces to

$$\frac{T\,d_iS}{dt} = J_v(\Delta P - RT\Delta c_i) + \frac{J_iRT\,\Delta c_i}{\dot{c}_i} + I\mathcal{E} \tag{4.43}$$

The corresponding phenomenological equations are

$$J_v = L_{vv}(\Delta P - RT\Delta c_i) + L_{vi}(RT\Delta c_i/\dot{c}_i) + L_{vI}\mathcal{E} \tag{4.44}$$

$$J_i = L_{iv}(\Delta P - RT\Delta c_i) + L_{ii}(RT\Delta c_i/\dot{c}_i) + L_{iI}\mathcal{E} \tag{4.45}$$

and

$$I = L_{Iv}(\Delta P - RT\Delta c_i) + L_{Ii}(RT\Delta c_i/\dot{c}_i) + L_{II}\mathcal{E} \tag{4.46}$$

where, once again, $L_{vi} = L_{iv}$, $L_{vI} = L_{Iv}$, and $L_{iI} = L_{Ii}$.

The problem now is to translate the straight and cross coefficients in equations (4.44) to (4.46) into experimentally useful or *practical* terms that define the properties of the system. This can be accomplished by examining the flows under experimental conditions when two of the three driving forces or flows are set equal to zero. Clearly:

(a) When $\Delta c_i = 0$ and $\mathcal{E} = 0$ (so that $\Delta\psi = 0$), equation (4.44) reduces to $J_v = L_{vv}\,\Delta P$, so that L_{vv} is obviously equal to L_p. Under the same conditions $J_i = L_{iv}\,\Delta P$, so that from equations (4.24) and (4.25) $L_{iv} = (1 - \sigma_i)\,c_i L_p = L_{vi}$.

(b) When $(\Delta P - RT\Delta c_i) = 0$ and $\mathcal{E} = 0$ but $\Delta c_i \neq 0$, then

$$J_i = L_{ii}(RT\Delta c_i/\dot{c}_i)$$

so that $L_{ii} = \omega_i \dot{c}_i$.

(c) When $\Delta P = 0$ and $\Delta c_i = 0$, then $I = L_{II}\mathcal{E} = L_{II}\Delta\psi$ so that L_{II} is simply the electrical conductance of the membrane G_m. Under the same conditions $J_v = L_{vI}\mathcal{E} = L_{vI}I/G_m$. Thus, we can define a new cross coefficient $\beta' = (L_{vI}/G_m)$, which is the volume flow associated with unit current flow when $\Delta P = 0$ and $\Delta c_i = 0$ or the *electro-osmotic* permeability. Finally, under these conditions $J_i = L_{iI}\mathcal{E} = L_{iI}I/G_m$. Since the current carried by J_+, which we recall is equal to J_i, is simply $\mathcal{F}J_i$, we can define the transference number (i.e., the fraction of the total current carried by J_+ when $\Delta P = 0$ and $\Delta c_i = 0$) as

$$t_+ = (\mathcal{F}J_i/I) = \mathcal{F}L_{iI}/G_m$$

Thus, we have converted the three straight and three cross coefficients into a new set of experimentally meaningful coefficients, where L_p, ω_i, and G_m are the corresponding straight coefficients and σ_i, t_+, and β' are the cross coefficients.

We can now rewrite equations (4.44) to (4.46) as follows:

$$J_v = L_p\Delta P - \sigma_i L_p RT\Delta c_i + \beta' I \tag{4.47}$$

$$J_i = L_p(1 - \sigma_i)\,\dot{c}_i(\Delta P - RT\Delta c_i) + \omega_i RT\Delta c_i + t_+ I/\mathcal{F} \tag{4.48}$$

and

$$I = \beta' G_m(\Delta P - RT\Delta c_i) + G_m t_+ RT\Delta c_i/\mathcal{F}\dot{c}_i + G_m\mathcal{E} \tag{4.49}$$

Now, although equations (4.47) to (4.49) appear somewhat formidable, they provide insight into a number of well-described electrokinetic phenomena. For example, if both bathing solutions have identical compositions so that $\Delta c_i = 0$ and $\mathcal{E} = \Delta\psi$, the equations describing volume and current flow are simply

$$J_v = L_p\Delta P + \beta' I \tag{4.50}$$

and

$$I = G_m(\beta'\Delta P + \Delta\psi) \tag{4.51}$$

Thus, when a current is passed across the membrane, a pressure must be applied by the piston to prevent electro-osmotic volume flow; this pressure, referred to as the *electro-osmotic pressure*, is simply.

$$P_{\mathcal{E}} = -\beta' I/L_p \tag{4.52}$$

If a pressure is applied across the membrane, producing a volume flow under zero current conditions (i.e., in the absence of an external current source), a *streaming potential* will be generated given by

$$-\Delta\psi = \beta'\Delta P = \beta' J_v/L_p \tag{4.53}$$

and the *streaming current* (i.e., the external current necessary to clamp $\Delta\psi$ to zero in the presence of volume flow) is

$$I = \beta' G_m \Delta P \tag{4.54}$$

Finally, by simply manipulating equations (4.50) and (4.51) the following identities emerge:

$$\text{(a)} \quad (J_v/\Delta\psi)_{\Delta P=0} = (I/\Delta P)_{\Delta\psi=0} \tag{4.55}$$

so that the relation between electro-osmotic volume flow and $\Delta\psi$ when there is no applied hydrostatic pressure is equal to the relation between the (streaming) current necessary to reduce $\Delta\psi$ to zero and an applied pressure.

$$\text{(b)} \quad (I/J_v)_{\Delta\psi=0} = -(\Delta P/\Delta\psi)_{J_v=0} \tag{4.56}$$

so that the relation between the streaming current and a pressure-induced volume flow is equal to the ratio of the electro-osmotic pressure necessary to prevent volume flow and the applied electrical potential difference.

These relations have been experimentally confirmed repeatedly and support the validity of applying Onsager's reciprocal relations to systems in which the flows of solvent and electrolytes are coupled (Miller, 1960).

Finally, it should be noted that in systems characterized by at least two flows and two forces, the straight and cross coefficients can be experimentally defined in several ways. For example, the electrical conductance of the membrane G_m can be determined under a variety of experimental conditions. Obviously, one desired condition is when $\Delta c_i = 0$ so that $\mathcal{E} = \Delta\psi$. However, even under this condition, we can determine G_m, or $\partial I/\partial(\Delta\psi)$, either when $J_v = 0$ *or* when $\Delta P = 0$. The values obtained will, in general, not be equal since

$$(G_m)_{\Delta P=0} = (G_m)_{J_v=0}/(1 + P'_{\mathcal{E}}\beta')$$

where $P'_{\mathcal{E}} = (\Delta P/\Delta\psi)$ when J_v and Δc_i are zero; $P'_{\mathcal{E}}$ is, therefore, a coefficient of electro-osmotic pressure. Thus, there is no "unique" electrical conductance, and the choice of values is a matter of experimental necessity or convenience.

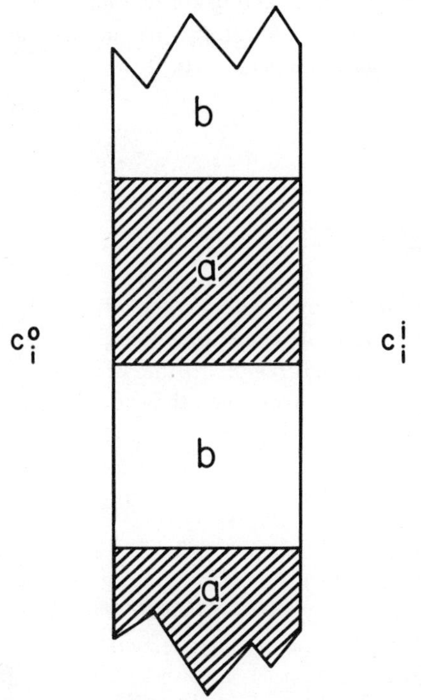

Figure 4.3. A composite membrane made up of different regions arranged in parallel.

The same principle applies to *all* of the phenomenological coefficients. The important point is that they are defined for a given set of experimental conditions, and their values can only be applied under the same set of conditions.

4.5. Transport of solute and solvent across composite membranes

Up to now we have been concerned with interactions between the flows of solute and solvent across homogeneous membranes that have uniform properties throughout and can be characterized by a single set of *elemental* straight and cross coefficients. We will now consider inhomogeneous or composite membranes that, because of interactions between flows, can exhibit properties not seen in simpler systems. We will limit our formal considerations to the flows of an uncharged solute and solvent and simply point out, where appropriate, the behavior that can be observed with charged species. The approach will closely follow that of Kedem and Katchalsky (1963*b*, *c*) and House (1974).

Let us first consider the membrane illustrated in Figure 4.3, which is made up of two regions *a* and *b* that have different properties and are arranged parallel to one another. In such a system the forces acting across each region are the same, and the total flow of any substance is simply the sum of the parallel flows across each region. Thus, we can write for the *j*th flow

$$J_j = \gamma^a J_j^a + \gamma^b J_j^b \tag{4.57}$$

where the γ's are fractional areas, and

$$J_j^a = \sum_{k=j}^{n} L_{jk}{}^a X_k \quad \text{and} \quad J_j^b = \sum_{k=j}^{n} L_{jk}{}^b X_k \tag{4.58}$$

where J_j is expressed per unit area occupied by regions *a* and *b*; clearly, $\gamma^a + \gamma^b = 1$. The set of *elemental* straight and cross phenomenologic coefficients $L_{jk}{}^a$ and $L_{jk}{}^b$ relate the flows across each region to the *same set* of driving forces; if the two regions have different properties, they must differ in one or more of these coefficients.

Now, the *total observable* flow is given by

$$J_j = \sum_{k=j}^{n} L_{jk} X_k \tag{4.59}$$

where the set of coefficients designated by L_{jk} are the *overall observable* coefficients and are some combination of the elemental coefficients. Substituting equations (4.58) and (4.59) into equation (4.57), we see that each overall coefficient is simply the area-weighted sum of the corresponding elemental coefficients, or

$$L_{jk} = \gamma^a L_{jk}{}^a + \gamma^b L_{jk}{}^b \tag{4.60}$$

This is not surprising in view of the fact that the *L*'s are generalized conductances, so that strict additivity is analogous to Kirchhoff's law of conductance in a parallel electrical circuit.

Now, the interacting flows of solute *i* and solvent *w* across region *a* can be described by equations (4.19) as follows:

$$J_v{}^a = L_p{}^a \Delta P + L_{pD}{}^a RT\Delta c_i \tag{4.61}$$

and

$$J_D{}^a = L_{pD}{}^a \Delta P + L_D{}^a RT\Delta c_i \tag{4.62}$$

and a similar set of equations can be written for region *b*.

It follows that

$$L_p = \gamma^a L_p{}^a + \gamma^b L_p{}^b \tag{4.63}$$

$$L_{pD} = \gamma^a L_{pD}{}^a + \gamma^b L_{pD}{}^b \tag{4.64}$$

and

$$L_D = \gamma^a L_D{}^a + \gamma^b L_D{}^b \tag{4.65}$$

Equation (4.63) simply states that the overall L_p is the area-weighted sum of the elemental L_p's; in view of the fact that L_p is a pure conductivity, this conclusion is not surprising.

Since $\sigma_i = -(L_{pD}/L_p)$, it follows that

$$\sigma_i = -(\gamma^a L_{pD}{}^a + \gamma^b L_{pD}{}^b)/L_p$$

or

$$\sigma_i = -(\gamma^a L_{pD}{}^a/L_p) - (\gamma^b L_{pD}{}^b/L_p)$$

Multiplying the first term on the right by $(L_p{}^a/L_p{}^a)$ and the second term by $(L_p{}^b/L_p{}^b)$, we obtain

$$\sigma_i = \sigma_i{}^a(\gamma^a L_p{}^a/L_p) + \sigma_i{}^b(\gamma^b L_p{}^b/L_p) \tag{4.66}$$

Now, when $\Delta c_i = 0$ and a pressure difference ΔP is applied across the membrane, $J_v = L_p\ \Delta P$, $J_v{}^a = L_p{}^a\ \Delta P$, and $J_v{}^b = L_p{}^b\ \Delta P$. Thus,

$$\left(\frac{\gamma^a L_p{}^a}{L_p}\right) = \left(\frac{\gamma^a J_v{}^a}{J_v}\right) \quad \text{and} \quad \left(\frac{\gamma^b L_p{}^b}{L_p}\right) = \left(\frac{\gamma^b J_v{}^b}{J_v}\right)$$

It follows that *the overall σ_i is the sum of the elemental reflection coefficients of the two regions weighted by the fractional volume flow through these regions in response to an applied pressure.*

Kedem and Katchalsky (1963b) have shown that for electroyltes there is an additional term in the expression for σ_i that may either be positive or negative, so that the overall reflection coefficient may be greater than or smaller than either of the elemental σ's.

As discussed in Section 4.3, $\omega_i = \dot{c}_i(L_p L_D - L_{pD}{}^2)/L_p$, so that using equations (4.63) to (4.65) it can be readily shown that

$$\omega_i = \gamma^a \omega_i{}^a + \gamma^b \omega_i{}^b + (\sigma_i{}^a - \sigma_i{}^b)^2 \left(\frac{\gamma^a \gamma^b L_p{}^a L_p{}^b}{L_p}\right) \dot{c}_i \tag{4.67}$$

Thus, unlike the cases for L_p and σ_i, ω_i is not simply an appropriately weighted combination of the elemental ω's except when $\sigma_i{}^a = \sigma_i{}^b$; if this equality does not hold, the overall ω_i will *always be greater* than the area-weighted sum of the elemental ω's.

What is the origin of this significant departure from the behavior of a homogeneous membrane? Recall that ω_i is defined as $(J_i/RT\Delta c_i)$ when $J_v = 0$. Thus, in order to determine ω_i, we must apply a hydrostatic pressure to compartment i of Figure 4.1 sufficient to abolish J_v (for the sake of this illustration, we will continue to assume that $c_i{}^i > c_i{}^o$). The required pressure is

$$\Delta P = \sigma_i RT\Delta c_i = \left[\left(\frac{\gamma^a L_p{}^a}{L_p}\right)\sigma_i{}^a + \left(\frac{\gamma^b L_p{}^b}{L_p}\right)\sigma_i{}^b\right] RT\Delta c_i \tag{4.68}$$

Under these conditions

$$J_v = \gamma^a J_v{}^a + \gamma^b J_v{}^b = 0 \tag{4.69}$$

Now, from equation (4.24)

$$\gamma^a J_v{}^a = \gamma^a L_p{}^a(\Delta P - \sigma_i{}^a RT\Delta c_i) \tag{4.70}$$

and

$$\gamma^b J_v{}^b = \gamma^b L_p{}^b(\Delta P - \sigma_i{}^b RT\Delta c_i) \tag{4.71}$$

Substituting equations (4.63) and (4.68) into equation (4.70) and (4.71), we obtain

$$\gamma^a J_v{}^a = -\left(\frac{\gamma^a L_p{}^a \gamma^b L_p{}^b}{L_p}\right)(\sigma^b - \sigma^a)RT\Delta c_i \tag{4.72}$$

and

$$\gamma^b J_v{}^b = \left(\frac{\gamma^a L_p{}^a \gamma^b L_p{}^b}{L_p}\right)(\sigma^b - \sigma^a)RT\Delta c_i \tag{4.73}$$

Thus, when $J_v = 0$ and $\sigma_i{}^a = \sigma_i{}^b$, there will be no volume flow across regions *a and b.* But when $\sigma_i{}^a \neq \sigma_i{}^b$, the condition of zero *overall* volume flow will be achieved by the *presence of equal and oppositely directed volume flow across the two different regions;* in other words, a *fluid circuit* will be established. Further, each of these fluid streams may be associated with a coupled solute flow so that there may also be a recirculation of solute across the membrane. The direction and relative solute content of each stream can be deduced intuitively. If $\sigma_i{}^b$ is greater than $\sigma_i{}^a$, the fluid stream through region *b* will be directed toward compartment *i* (i.e., toward the compartment with the higher concentration), but will be relatively poor with respect to solute content. On the other hand, a fluid stream of equal volume will flow from compartment *i* to compartment *o* through region *a*, which will be richer in solute content. Thus, the difference in solute contents of the oppositely directed fluid streams contributes to the flow of *i* from compartment *i* to compartment *o* (which is assumed to be strictly diffusional when $J_v = 0$). The total effect is to *augment* the rate of appearance of *i* in compartment *o* and, thereby, to *exaggerate* the apparent permeability of the membrane to *i*.

This phenomenon can be described formally by writing the appropriate equations for solute flow, that is,

$$\gamma^a J_i^a = \gamma^a J_v{}^a(1 - \sigma_i{}^a)\bar{c}_i + \gamma^a \omega_i{}^a RT\Delta c_i \tag{4.74}$$

and

$$\gamma^b J_i^b = \gamma^b J_v{}^b(1 - \sigma_i{}^b)\bar{c}_i + \gamma^b \omega_i{}^b RT\Delta c_i \tag{4.75}$$

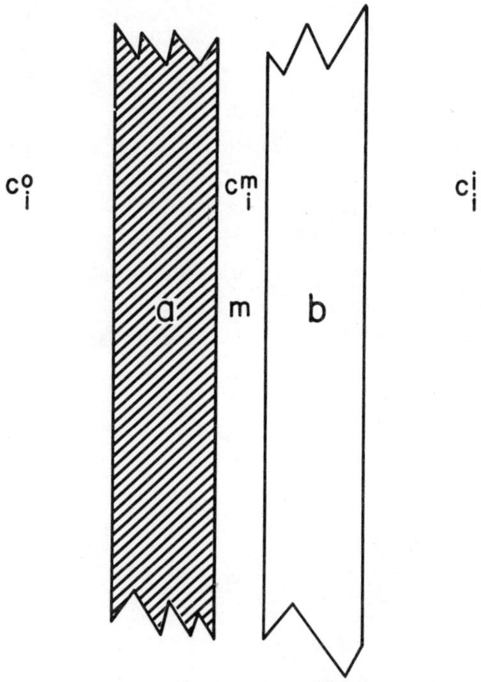

Figure 4.4. A composite barrier made up of two different regions arranged in series separated by an intermediate zone that can be considered infinitesimally thin.

Substituting $\gamma^b J_v{}^b = -\gamma^a J_v{}^a$ into equation (4.75) and summing equations (4.74) and (4.75) yields

$$J_i = \gamma^a J_v{}^a(\sigma_i{}^b - \sigma_i{}^a)\bar{c}_i + (\gamma^a \omega_i{}^a + \gamma^b \omega_i{}^b)RT\Delta c_i \tag{4.76}$$

Clearly, if $\sigma_i{}^b > \sigma_i{}^a$, $J_v{}^a$ is negative (directed from compartment i to compartment o) and J_i is augmented by the differential solute flow.[4] Alternatively, if $\sigma_i{}^b < \sigma_i{}^a$, then $J_v{}^a$ is positive (i.e., it is directed from compartment o to compartment i) and, once again, J_i is augmented. From equation (4.76) it is clear that (when $J_v = 0$)

$$(J_i/RT\Delta c_i) = \omega_i = (\gamma^a \omega_i{}^a + \gamma^b \omega_i{}^b)$$

only when $\sigma_i{}^a = \sigma_i{}^b$!

Let us now turn our attention to a composite membrane made up of two different barriers arranged in series separated by an infinitely thin intermediate region m as shown in Figure 4.4. The relations between flows and forces that define such an array are

$$X_j = X_j{}^a + X_j{}^b \tag{4.77}$$

where $X_j^a = X_j^o - X_j^m$ and $X_j^b = X_j^m - X_j^i$, and

$$J_j = \sum_{k=j}^{n} L_{jk}^a X_k^a = \sum_{k=j}^{n} L_{jk}^b X_k^b \tag{4.78}$$

Equation (4.77) asserts that the *total* jth force across the system is the sum of the forces across barrier a and barrier b and is simply a statement of the principle that the difference in intensive properties of state between two points is independent of path. Equation (4.78) asserts that under steady-state conditions the flows across each barrier arranged in series are equal; this is simply a statement of the conservation of matter.

By definition, ω_i is given by $(J_i/RT\Delta c_i)$ when $J_v = 0$. Under these conditions, according to equation (4.78),

$$J_i = \omega_i^a RT\Delta c_i^a = \omega_i^b RT\Delta c_i^b = \omega_i RT\Delta c_i \tag{4.79}$$

where $\Delta c_i^a = c_i^o - c_i^m$ and $\Delta c_i^b = c_i^m - c_i^i$ so that $\Delta c_i = \Delta c_i^a + \Delta c_i^b$. It follows that

$$J_i[(1/\omega_i^a) + (1/\omega_i^b)] = RT\Delta c_i$$

so that

$$(1/\omega_i) = (1/\omega_i^a) + (1/\omega_i^b) \tag{4.80}$$

Since $(1/\omega_i)$ is simply the overall resistance of the barrier to solute diffusion, equation (4.80) is a statement of the strict additivity of resistances arranged in series.

Similarly, when $J_v = 0$, $\Delta P = \sigma_i RT\Delta c_i$ and the analogous relations across the two membranes are $\Delta P^a = \sigma_i^a RT\Delta c_i^a$ and $\Delta P^b = \sigma_i^b RT\Delta c_i^b$. From these equalities and those given in equation (4.79) it follows that

$$\sigma_i = \sigma_i^a(\omega_i/\omega_i^a) + \sigma^b(\omega_i/\omega_i^b) \tag{4.81}$$

Thus, the overall reflection coefficient is the sum of the two elemental coefficients weighted by the relative resistances of the two barriers to the diffusional flow of solute.

Equations (4.80) and (4.81) have several interesting implications.

First, if either ω_i^a or ω_i^b is dependent upon concentration, the system illustrated in Figure 4.4 will exhibit an asymmetry with respect to "permeability." That is, the ω_i measured with concentration c_i^o in compartment o and c_i^i in compartment i will differ from that measured when the solutions are reversed. Further, under these conditions σ_i will also display a polarity [equation (4.81)], so that the effective osmotic pressure across such a barrier for a given Δc_i will depend upon the orientation of the solutions.

But, perhaps the most interesting and unexpected property of the series array illustrated in Figure 4.4 is the relation between the overall L_p and the elemental L_p's. Since L_p is given by $(J_v/\Delta P)$ when $\Delta c_i = 0$, we can write

$$J_v = L_p{}^a(\Delta P^a - \sigma_i{}^a RT\Delta c_i{}^a) = L_p{}^a(\Delta P^b - \sigma_i{}^b RT\Delta c_i{}^b) \qquad (4.82)$$

where, according to equation (4.77)

$$\Delta P = \Delta P^a + \Delta P^b \qquad (4.83)$$

Since when $\Delta c_i = 0$

$$RT\Delta c_i{}^a = -RT\Delta c_i{}^b \qquad (4.84)$$

it follows that

$$J_v\left(\frac{L_p{}^a + L_p{}^b}{L_p{}^a L_p{}^b}\right) = \Delta P - (\sigma_i{}^a - \sigma_i{}^b)RT\Delta c_i{}^a \qquad (4.85)$$

Thus, in spite of the fact that the L_p's are conductances, the overall L_p is *not* simply equal to $[L_p{}^a L_p{}^b/(L_p{}^a + L_p{}^b)]$ as would be expected for the summation of electrical conductances in series *unless* $\sigma_i{}^a = \sigma_i{}^b$ *or* $\Delta c_i{}^a = 0$.

Now, $\Delta c_i{}^a$ in equation (4.85) can be evaluated from the following relations:

$$J_i = (1 - \sigma_i{}^a)J_v\dot{c}_i{}^a + \omega_i{}^a RT\Delta c_i{}^a = (1 - \sigma_i{}^b)J_v\dot{c}_i{}^b + \omega_i{}^b RT\Delta c_i{}^b \qquad (4.86)$$

where $\dot{c}_i{}^a$ and $\dot{c}_i{}^b$ are the average concentrations across barriers a and b, respectively, given by equation (4.14). Clearly, when $\Delta c_i = 0$, $\dot{c}_i{}^a = \dot{c}_i{}^b$; thus, from equations (4.84) and (4.86) it follows that

$$RT\Delta c_i{}^a = \dot{c}_i{}^a\left(\frac{\sigma_i{}^a + \sigma_i{}^b}{\omega_i{}^a + \omega_i{}^b}\right)J_v \qquad (4.87)$$

Substituting equation (4.87) into (4.85) and rearranging, we obtain

$$J_v\left\{\left(\frac{L_p{}^a + L_p{}^b}{L_p{}^a L_p{}^b}\right) + \left[\frac{(\sigma_i{}^a - \sigma_i{}^b)^2}{\omega_i{}^a + \omega_i{}^b}\right]\dot{c}_i{}^a\right\} = \Delta P \qquad (4.88)$$

Thus,

$$\left(\frac{1}{L_p}\right) = \left(\frac{1}{L_p{}^a}\right) + \left(\frac{1}{L_p{}^b}\right) + \left[\frac{(\sigma_i{}^a - \sigma_i{}^b)^2}{\omega_i{}^a + \omega_i{}^b}\right]\dot{c}_i{}^a \qquad (4.89)$$

According to equation (4.89), the resistance to volume flow across this composite membrane will *always exceed* the sum of the resistances of the two series barriers unless $\sigma_i{}^a = \sigma_i{}^b$.

The physical reason for this intuitively unexpected property arises from the behavior of the central compartment. Thus, let us assume that $\Delta c_i = 0$ and that a positive pressure is suddenly applied to compartment i. If $\sigma_i{}^b < \sigma_i{}^a$, the fluid stream crossing barrier b from compartment i to compartment m will contain more solute than the stream that flows from compartment m to compartment o across barrier a. Consequently, solute will accumulate in the intermediate compartment. In time, a steady state will be achieved in which $c_i^m > c_i^o$ and the effective pressure generated by the concentration difference (Δc_i^a) across the less permeable (the larger σ_i) barrier a will oppose ΔP and retard volume flow across the *entire* system. Alternatively, if $\sigma_i^b > \sigma_i^a$, the sudden application of a hydrostatic pressure to compartment i will deplete solute from the central compartment, since the solute content of the fluid stream crossing membrane b will be less than that contained by the stream crossing membrane a. Thus, when a steady state is achieved, $c_i^m < c_i^i$ and the effective osmotic pressure across membrane b will oppose volume flow in response to the externally applied hydrostatic pressure. In short, if $\sigma_i^a \neq \sigma_i^b$, there will be an accumulation or a depletion of solute in the central compartment that, in either instance, will lead to the development of an "intramembranous effective osmotic pressure" that reduces the effect of ΔP on J_v. Thus, the observed overall resistance to volume flow will always exceed the sum of the two elemental resistances. Further, as discussed by Kedem and Katchalsky (1963*c*) the degree of solute depletion or accumulation in the central compartment is a function of the direction and magnitude of J_v so that, in general, a plot of J_v versus ΔP will be nonlinear and will display rectification (i.e., for a given pressure difference the resistance to flow will be greater in one direction than the other); this point is illustrated by House (1974, p. 74).

4.6. Summary

Although the irreversible (or nonequilibrium) thermodynamic approach does not enjoy the unassailable status of classical thermodynamics, it has proved to be remarkably successful for the description of interactions among flows and forces in relatively well-defined membrane systems. The applicability of the linear phenomenological equations and Onsager's relations have been repeatedly demonstrated for the cases of solute-solvent and solute-solute interactions even in systems that are significantly displaced from equilibrium. The descriptions are phenomenologic and do not rest on assumptions regarding mechanism. Thus, although the approach can provide an internally consistent and experimentally useful set of descriptive parameters, it sheds little insight into underlying mechanism. As noted, some efforts have been made to translate the phenomenological coefficients into more mechanistically meaningful terms, but it is not always clear whether these represent real advances or whether they simply substitute one set of phenomenologic statements with another. For this reason, the approach has been subjected to some mixtures of skepticism and criticism. Clearly, a great deal more must be learned about the microscopic properties of solvent, solutes, and membranes before realistic, mechanistic models of transport processes can be formulated; but, until then, a useful and internally consistent phenomenologic treatment is better than none at all. Certainly, the contributions of classical thermodynamics, which rank among the most noble achievements of the human intellect, cannot be criticized as being devoid of meaning simply because many antedated our current concepts regarding the molecular structure of matter.

Finally, there are a number of important practical messages to be gleaned from the phenomenologic treatment of composite membranes.

First, we see that asymmetric behavior (e.g., rectification of solute permeability coefficients, reflection coefficients, and pressure-volume flow relations) which, on intuitive grounds, is not generally expected of an inert barrier can arise simply from the presence of heterogeneities along the breadth or thickness of the barrier. Clearly, an inert membrane penetrated by uniform pores having different radii and/or by nonuniform pores that are narrow at one end and wider at the other would be expected to display such behavior and no more complex explanations need be invoked.

Second, it is clear from these simple considerations that *overall* straight (L_p and ω_i) and cross coefficients (σ_i) need not bear a clear-cut relation to the respective elemental coefficients (i.e., they need not simply be area-weighted averages). Thus "equivalent" pore dimensions of complex membranes calculated from values of L_p using Poiseuille's law or from values of σ_i using the Renkin-Durbin approach (see Section 4.3) may be entirely unrealistic.

5 Active transport

5.1. A definition of active transport

As discussed briefly in Chapter 1, in a system characterized by a single force and a single flow, the flow *must* be directed from a region of higher electrochemical potential to one of lower electrochemical potential so that $T\,d_iS/dt = J_i X_i > 0$. In other words the flow and the force must have the same sign. However, if there are two or more flows and forces, *negative coupling* between flows can drive a substance in a direction opposite to its conjugate force; we referred to this phenomenon as *energy conversion* (see Section 1.6). Such phenomena are, of course, characteristic of biological membranes, but can also be observed in artificial membrane systems. For example, as discussed in Section 4.3, if a membrane is more selective for solute than solvent ($\sigma_i < 0$), solvent flow will be directed from a region of high solute concentration to the region of lower solute concentration, a direction opposite to the osmotic driving force (*anomalous osmosis*). Further, from equation (4.25) we see that J_i can be negative if J_v is negative so that interactions between the flows of i and solvent can drive i from the compartment of lower concentration to that of higher concentration (e.g., *solvent drag*).

In 1948 Rosenberg proposed that the term *active transport* should be applied to flows that are directed against their conjugate driving forces; that is, J_i is "active" if $J_i X_i < 0$. Conversely, if $J_i X_i > 0$, J_i is considered "passive." This is an unambiguous definition based entirely on classical thermodynamic grounds that can be readily applied experimentally.

An alternate definition was proposed by Kedem (1961), who simply rearranged equation (1.35) as follows:

$$J_i = (X_i/R_{ii}) - \sum_j (R_{ij}J_j/R_{ii}) - (R_{ir}J_r/R_{ii}) \tag{5.1}$$

where J_r is the flow of a chemical reaction and R_{ir} is the cross coefficient that quantitates the coupling between J_i and J_r. The J_j represents *all other* flows. Thus, in an isothermal system J_i is determined by its conjugate driving force, interactions with the flows of solvent and/or other solutes ("osmo-osmotic coupling") and coupling to the flow of a chemical reaction ("chemi-osmotic coupling"). Kedem suggested that the term "active transport" be reserved for those flows of i where $R_{ir} \neq 0$. That is, J_i is "active" if it is coupled to a chemical reaction; *if not*, it is referred to as "passive."

The definition proposed by Kedem has a number of advantages over that proposed by Rosenberg (Curran and Schultz, 1968; Schultz, 1968).

First, it serves to distinguish between flows that are directed against their conjugate driving forces because of interactions with the flows of other solutes or solvent and those that are driven by a chemical reaction. Some refer to the former as "secondary active (osmo-osmotic) transport" and to the latter as "primary active (chemi-osmotic) transport" (Mitchell 1967, 1970). Since a major purpose of any categorization of transport processes is to direct future research, a definition that implies coupling to a biochemical reaction is more useful than one based exclusively on the direction of the transport process.

Second, the Kedem definition does not depend upon the direction of flow. The direction of J_i may be the same as that of X_i but it still will be considered *active* if it is accelerated or slowed by "positive" or "negative" coupling to J_r. In contrast, the Rosenberg definition, which hinges on the direction of J_i, can lead to an awkward situation. For example, in the absence of coupling to the flows of J_j, equation (5.1) reduces to

$$J_i = (X_i/R_{ii}) - (R_{ir}J_r/R_{ii}) \tag{5.2}$$

Now, if R_{ii} is very large (e.g., the barrier is essentially impermeable to i), $J_i \cong -(R_{ir}J_r/R_{ii})$ and is essentially independent of X_i. Thus, in principle we could adjust X_i so that in one instance the flow of i is directed in the same direction as its conjugate force and in another *the same flow* is opposed to its driving force. According to the Rosenberg definition, in the first instance J_i would be considered passive and in the other it is considered active in spite of the fact that J_i is not influenced by X_i *and in both instances it is driven by the same mechanism!*

Thus, the Kedem definition departs from the classical thermodynamic grounds of the Rosenberg definition and introduces considerations of coupled flows and flow *rates.* However, although it is unambiguous and intuitively more satisfying than the Rosenberg definition, it is far more difficult to apply in practice. In principle, in order to unequivocally establish positive or negative coupling between J_i and a chemical (metabolic) reaction, either (a) the reaction must be identified and a nonzero R_{ir} must be demonstrated or (b) all possible J_j and R_{ij} must be determined in addition to R_{ii} and X_i; clearly, in a complex biological system this could pose formidable experimental obstacles.

5.2. Experimental criteria for categorizing transport processes

As discussed in Section 2.5, if the movement of a substance across a membrane is the result of simple diffusion and can be described by an integrated form of the Nernst–Planck equation, the relation between the two unidirectional fluxes is given by the Ussing flux-ratio equation

$$J_i^{oi}/J_i^{io} = (c_i^o/c_i^i)\exp(-z_i \mathcal{F}\Delta\psi/RT) \tag{5.3}$$

Since the advent of commercially available radioactive isotopes for many biologically important ions, this relation has been applied extensively and has provided the most important criterion for identifying the simplest of all transport processes. Thus, if the unidirectional fluxes conform to equation (5.3) over a wide range of values of c_i^o, c_i^i, and $\Delta\psi$, it is safe to conclude that the flow of i is driven solely by differences in chemical and electrical potentials and traverses the membrane in its free form without interactions, complex formation, and so on (see Section 2.5 for the assumptions underlying the flux-ratio equation).

It is an all-too-common misapprehension that the flux-ratio equation can be employed to distinguish between "active" and "passive" transport perhaps, in part, because of the title of this landmark contribution, "The distinction by means of tracers between active transport and diffusion" (Ussing, 1949). It must be emphasized that failure to conform with equation (5.3) does not exclude the possibility that the flows are driven entirely by their conjugate driving forces and most certainly does not imply the presence of coupling between the flow of i and a chemical reaction. In short, equation (5.3) simply describes a *sufficient but not necessary* criterion for the conclusion that the movements of a given substance are attributable solely to differences in electrical and chemical potentials.

It is instructive to consider, briefly, processes that can lead to deviations from the flux-ratio equation inasmuch as some of these departures can be distinctive. For the sake of simplicity we will consider an uncharged species or an ion when $\Delta\psi$ is clamped to zero. Under these conditions equation (5.3) reduces to

$$(J_i^{oi}/J_i^{io}) = (c_i^o/c_i^i)$$

Single-file diffusion

Suppose i crosses the membrane through narrow pores that can only accommodate a single row of two or more particles. Under these conditions tracer and parent species will not cross the barrier randomly. Instead, the tracer movement will interact with nontracer movement, since it must pass through the pore by a series of jumps caused by the impact of collisions at the two ends of the pore.

It can be readily shown (Hodgkin and Keynes, 1955) that this will accelerate the movement of the tracer in the direction of its electrochemical potential difference and retard the flow in the opposite direction. The flux-ratio equation takes the form

$$J_i^{oi}/J_i^{io} = [(c_i^o/c_i^i) \exp(-z_i \mathcal{F}\Delta\psi/RT)]^n \tag{5.4}$$

where n is the number of particles in the single file. When $z_i = 0$ or $\Delta\psi = 0$, equation (5.4) reduces to $(J_i^{oi}/J_i^{io}) = (c_i^o/c_i^i)^n$. Clearly, equation (5.2) will conform with the flux-ratio equation (5.3) only when

$$[(c_i^o/c_i^i) \exp(-z_i \mathcal{F}\Delta\psi/RT)] = 1$$

or $J_i = 0$. (For other formal descriptions of single-file diffusion see Sjodin, 1965; Hladky, 1965; Heckmann, 1972; and Hille and Schwarz, 1978.)

Membrane interactions

If the movement of an uncharged solute across a membrane is the result of complex formation with a mobile membrane component, or "carrier," but is not affected by any nonconjugate driving force, the process is referred to as *facilitated transfer.* This "carrier-mediated" process will be discussed in greater detail in Section 6.2. Suffice it to say at this point that the flux ratio is given by

$$(J_i^{oi}/J_i^{io}) = [c_i^o(K_t + c_i^i)]/[c_i^i(K_t + c_i^o)] \tag{5.5}$$

where K_t may be a complex parameter and is defined by equation (6.10). Clearly, equation (5.5) will only conform to the Ussing flux-ratio equation under the trivial condition when $c_i^o = c_i^i$ and $J_i = 0$.

Finally, significant deviations from the Ussing flux-ratio equation can result if, in addition to simple diffusion, the molecule interacts with a membrane component that mediates an obligatory one-for-one exchange of the solute across the barrier. This type of transfer process is termed *exchange diffusion* and was first proposed by Levi and Ussing (1948). It will be discussed in greater detail in Section 6.2; suffice it to say at this point that it closely resembles facilitated transfer, except that it cannot give rise to a net flux but simply increments both unidirectional fluxes equally. If simple diffusion is complicated by exchange diffusion, the flux ratio for an uncharged solute is given by

$$\frac{J_i^{oi}}{J_i^{io}} = \frac{c_i^o + (J_i^m/P_i)}{c_i^i + (J_i^m/P_i)} \tag{5.6}$$

where J_i^m is the maximum velocity of exchange diffusion (see Section 6.2). Clearly, the numerator and denominator of the right-hand expression are incremented equally by a term that weights the magnitude of the diffusional process and the contribution from the exchange process. When $J_i^m \gg P_i$, then (J_i^{oi}/J_i^{oi}) approaches unity; when $P_i \gg J_i^m$, the contribution of the exchange process is small and $(J_i^{oi}/J_i^{io}) \rightarrow (c_i^o/c_i^i)$.

Interaction between tracer and parent species

One of the assumptions underlying the derivation of the flux-ratio equation is that the flows of the tracer and the parent species do not interact (i.e., the cross coefficients are zero, so that each flow is driven exclusively by its conjugate driving force). In general, this may not be true, and Kedem and Essig (1965) and Essig (1966) have described the effects of positive and negative coupling between these flows on the flux ratio. An important consequence of their analysis is that these interactions can give rise to departures from the predicted ratio that resemble those attributed to single-file diffusion and exchange diffusion. These results caution against the inference of specific membrane properties or transport mechanisms on the basis of departures from the predicted flux ratios.

It should be stressed that in all of the conditions described here, which can lead to significant deviations from the flux-ratio equation, J_i is driven entirely by X_i and thus must be considered "passive" by *any* criterion. Let us now consider situations where J_i is influenced by nonconjugate forces.

Solvent drag

As we did in Section 2.5, let us assume that compartment o contains a tracer j and that $c_j^i = 0$. According to equation (4.25), the net flux of j from compartment o to compartment i, which is equivalent to the unidirectional flux, is given by

$$J_j^{oi} = (c_j^o/2)\,[2\omega_j RT + (1 - \sigma_j)\,J_v] \tag{5.7}$$

Similarly, if compartment i contains a tracer k and $c_k^o = 0$

$$J_k^{io} = (c_k^i/2)\,[2\omega_k RT - (1 - \sigma_k)\,J_v] \tag{5.8}$$

Assuming that j and k are perfect tracers for i, the flux ratio is given by

$$\frac{J_i^{oi}}{J_i^{io}} = \left(\frac{c_i^o}{c_i^i}\right) \cdot \left[\frac{2\omega_i RT + (1 - \sigma_i)\,J_v}{2\omega_i RT - (1 - \sigma_i)\,J_v}\right] \tag{5.9}$$

Thus, unless $J_v = 0$ or $\sigma_i = 1$, one unidirectional flux will be augmented and the other will be retarded by volume flow. The magnitude of this effect clearly depends on the relative magnitudes of $\omega_i RT$ and $(1 - \sigma_i)\,J_v$.

A similar expression can be written for possible solute-solute interactions (*solute drag*), which takes the form

$$\frac{J_i^{oi}}{J_i^{io}} = \left(\frac{c_i^o}{c_i^i}\right) \left[\frac{2\omega_i RT + L_{ji} J_j}{2\omega_i RT - L_{ji} J_j}\right] \tag{5.10}$$

where J_j is the net flow of a species j and L_{ji} is a measure of the extent of coupling between J_j and the flows of i.

Active transport

The effect of superimposing an active transport process on a simple diffusional process depends on the characteristics of the former. In the simplest case if i is actively transported from compartment o to compartment i (see Section 6.5)

$$J_i^{oi} = P_i c_i^o + [c_i^o J_i^m/(K_t + c_i^o)] \tag{5.11}$$

and $J_i^{io} = P_i c_i^i$. Thus, the flux ratio is given by

$$\frac{J_i^{oi}}{J_i^{io}} = \left(\frac{c_i^o}{c_i^i}\right)\left[1 + \frac{J_i^m}{P_i(K_t + c_i^o)}\right] \tag{5.12}$$

Summary

Figure 5.1 shows the relations between (J_i^{oi}/J_i^{io}) and (c_i^o/c_i^i) calculated from equations (5.3)-(5.6), (5.9), and (5.12). The value of c_i^i was fixed at 100 mM and c_i^o was varied from 20 to 200 mM; other transport parameters were selected arbitrarily for scaling purposes. We see that the deviations from the line of identity are in some instances quite distinctive. It is particularly important to note that for any value of (c_i^o/c_i^i) other than unity all of the systems we have considered yield flux ratios that could be misinterpreted as indicative of active transport. Thus, flux ratios must be interpreted with caution and can be particularly misleading when examined over a narrow range of concentrations. On the other hand, data obtained over a wide range of concentration ratios may provide clues to possible underlying transport processes.

5.3. An irreversible thermodynamic model for active transport

The Kedem definition of active transport implies that the "flow of a chemical reaction" J_r can be treated as is any other flow in a system and that coupling between this flow and the flow of matter can be viewed as the modus operandi of an active transport process. Accordingly, in a system in which i is actively transported as a result of coupling to J_r, we may write the dissipation function

$$T(d_iS/dt) = J_iX_i + J_rX_r \tag{5.13}$$

where X_r is the conjugate driving force for the flow J_r.

Now, what is the conjugate driving force for the flow of a chemical reaction? Consider the reaction

$$\nu_A A + \nu_B B + \cdots \longrightarrow \nu_C C + \nu_D D + \cdots \tag{5.14}$$

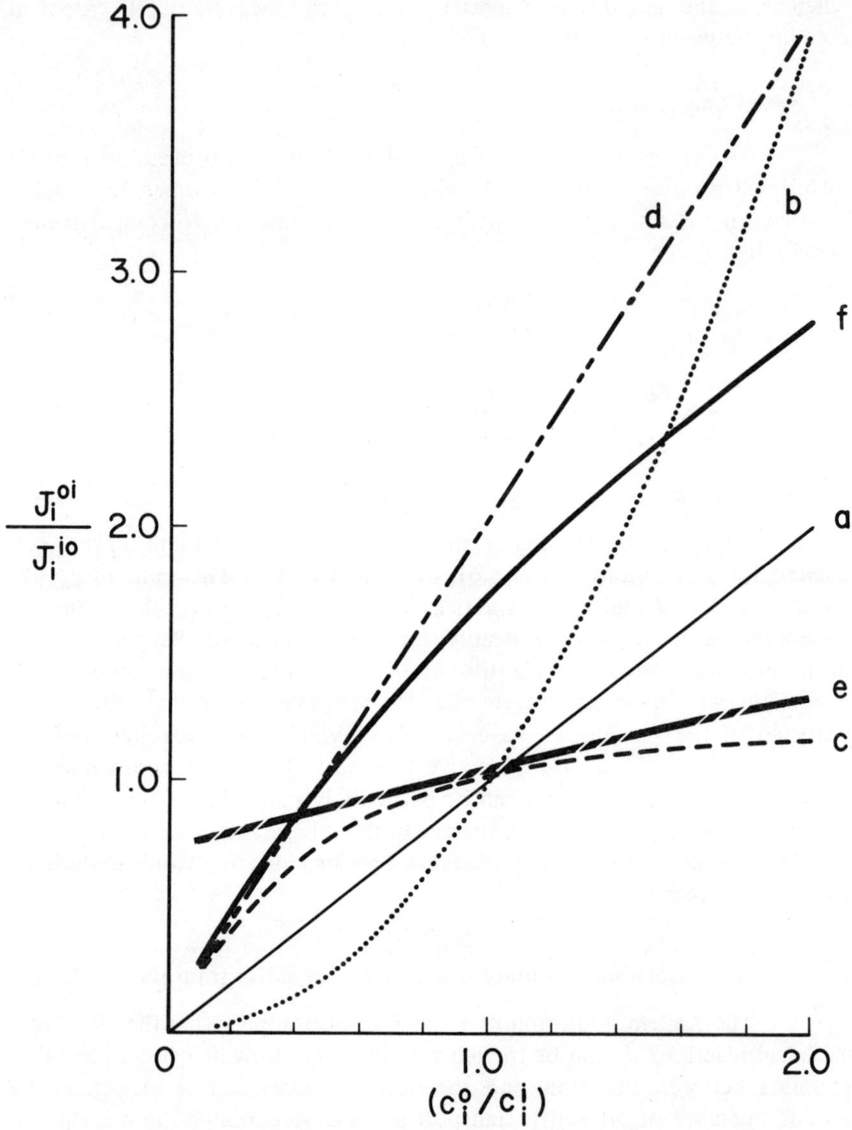

Figure 5.1. Relations between flux ratio and concentration ratio for an uncharged solute, determined as discussed in the text: *a*, simple diffusion; *b*, single-file diffusion; *c*, facilitated transfer; *d*, simple diffusion plus solvent drag; *e*, simple diffusion plus exchange diffusion; *f*, simple diffusion plus active transport.

where the ν's are stoichiometric coefficients of the reaction in which the reactants $A, B, \ldots$ are converted to the products $C, D, \ldots$ Assuming that only reactants are present at the beginning of the reaction, then, at any time thereafter the progress, extent, or *degree of advancement* of the reaction $d\zeta$ is given by the number of moles of reactant that have disappeared and/or the number of moles of product that have appeared. Accordingly,

$$d\zeta = -(dn_A/\nu_A) = -(dn_B/\nu_B) = (dn_C/\nu_C) = (dn_D/\nu_D) \cdots \tag{5.15}$$

Now, at constant T and P, the change in free energy of the reaction is

$$(dG)_{T,P} = \mu_A \, dn_A + \mu_B \, dn_B + \mu_C \, dn_C + \mu_D \, dn_D + \cdots \tag{5.16}$$

or

$$(dG)_{T,P} = \sum_j \mu_j \, dn_j \tag{5.17}$$

where dn_j is negative if j is a reactant and positive if j is a product. Substituting equation (5.15) into (5.17), we obtain

$$(dG)_{T,P} = d\zeta \sum_j \nu_j \mu_j \tag{5.18}$$

and differentiating by time (t), we obtain

$$(dG/dt)_{T,P} = (d\zeta/dt) \sum_j \nu_j \mu_j \tag{5.19}$$

where, now, the stoichiometric coefficients are negative for reactants and positive for products. Clearly, $d\zeta/dt$ is the *rate of advancement* of the reaction or the flow J_r. Thus, equation (5.19) can be rewritten

$$(dG/dt)_{T,P} = J_r \cdot \sum_j \nu_j \mu_j \tag{5.20}$$

As discussed in Section 1.3, the free energy of a system decreases during any spontaneous process (i.e., $dG < 0$). Further, as illustrated by equation (1.12), in a system at constant T and P that does not perform "useful work" $(-dG/dt) = T(d_iS/dt)$. Thus, equation (5.20) reduces to

$$T \, d_iS/dt = J_r \sum_j (-\nu_j \mu_j) \tag{5.21}$$

It follows that the conjugate driving force of the reaction is given by $\Sigma(-\nu_j \mu_j)$ which (at constant T and P) is the decrease in Gibbs free energy of the reaction per unit rate of advancement. This relation was formulated in 1936 by the pioneering work of DeDonder, who defined the driving force for the flow of a chemical reaction as the *affinity A*.

Assuming that flows and forces are linearly related, we can describe this two-force, two-flow system by the following equations (see equations 1.35):

$$X_i = \Delta\tilde{\mu}_i = R_{ii}J_i + R_{ir}J_r$$

and (5.22)

$$A = R_{ri}J_i + R_{rr}J_r$$

where the R's are generalized resistances and, assuming that the reciprocal relations apply, $R_{ir} = R_{ri}$.

As discussed in Section 1.6, in principle, the linear phenomenological equations and Onsager's relations are expected to apply only when a system is not displaced *too far* from equilibrium, and in the final analysis the validity of these relations must be established empirically. Whether biological transport processes can be described by means of linear relations among J_i, J_r, $\Delta\tilde{\mu}_i$, and A is an unsettled question. Katchalsky and Curran (1965) have argued, on the basis of kinetic analysis, that a linear relation between J_r and A is likely to hold only over a very limited range where the reaction is very close to equilibrium. Prigogine (1961) has also argued that linearity is to be expected only when $A < RT$ (approximately 600 cal/mole), a value that is considerably smaller than the free energy changes associated with most exergonic biochemical reactions. More recently, Prigogine and Lefever (1975) have shown that if an *overall* chemical reaction is the result of a number of intermediate reactions, each of which is not very far displaced from equilibrium, the *overall* reaction rate may be a linear function of the *overall* affinity even though the latter may be much larger than RT.

Experimental data bearing on this issue are limited. Blumenthal, Caplan, and Kedem (1967) have shown that the linear phenomenological equations and Onsager's reciprocal relations can be used to describe coupling between a chemical reaction and current flow (ion transport) in an artificial membrane system where A was as high as 3000 cal/mole. It is of interest that in their system (as in most enzymatically catalyzed reactions) J_r was a decidedly nonlinear function of A; nevertheless, the linear formalism yielded the correct data.

In short, the applicability of equations (5.22) to the description of biological transport processes is a moot point that cannot be settled on the basis of the available data. Nonetheless, it is instructive to carry through an analysis of these equations inasmuch as the results have bearing on energy converters in general and may provide some insight into biological transport processes. The approach we will employ closely follows that of Kedem and Caplan (1965) and Essig and Caplan (1968).

Rearranging equation (5.22), we obtain

$$J_i = \left(\frac{\Delta\tilde{\mu}_i}{R_{ii}}\right) - \left(\frac{R_{ir}J_r}{R_{ii}}\right) \quad (5.23)$$

We will focus mainly on the situation where coupling to J_r retards the flow of J_i in the direction of its conjugate driving force *or* actually drives J_i in the opposite direction so that $J_i\,\Delta\tilde{\mu}_i < 0$. This is referred to as *negative coupling* and if we define the directions of J_r and $\Delta\tilde{\mu}_i$ as positive, then $R_{ir} > 0$.

We start by defining two combinations of the phenomenological coefficients

$$Z = \left(\frac{R_{rr}}{R_{ii}}\right)^{1/2} \quad \text{and} \quad q = \frac{-R_{ir}}{(R_{ii}R_{rr})^{1/2}} \tag{5.24}$$

Since $R_{ir}{}^2 \leqslant R_{ii}R_{rr}$, it follows that $-1 \leqslant q \leqslant 1$; q is referred to as the *degree of coupling* for the following reasons:

(a) If $q = \pm 1$, then $(J_i/J_r) = \pm Z$. Thus, the ratio of the flows is independent of the ratio of the forces, and if one flow is fixed the other is uniquely determined; in other words, the two flows are completely (tightly) coupled. It follows that if the driving flow J_r is fixed then J_i is independent of $\Delta\tilde{\mu}_i$ and the system operates as a "constant current source." This is, of course, an ideal extreme. Alternatively, if J_i is affected by varying $\Delta\tilde{\mu}_i$, then J_r would be affected proportionally.

(b) If $q = 0$, then $R_{ir} = 0$, and

$$\left(\frac{J_i}{J_r}\right) = \frac{Z^2\Delta\tilde{\mu}_i}{A} = \frac{(\Delta\tilde{\mu}_i/R_{ii})}{(A/R_{rr})} = \frac{L_{ii}\Delta\tilde{\mu}_i}{L_{rr}A} \tag{5.25}$$

Thus, when $q = 0$, the ratio of flows is proportional to the ratio of forces; each flow is influenced only by its conjugate driving force so that the flows are entirely uncoupled.

(c) For values of q between 0 and -1 (negative coupling) the relation between the flow ratio and the force ratio varies as illustrated in Figure 5.2.

The quadrant delineated by the origin (0), the value $(J_i/J_rZ) = -1$, and the value $(\Delta\tilde{\mu}_iZ/A) = +1$ is referred to as the *driving region* since in this region $R_{ir}J_r > \Delta\tilde{\mu}_i$. Thus, J_i is *driven* in a direction opposite to $\Delta\tilde{\mu}_i$ (i.e., $J_i\Delta\tilde{\mu}_i < 0$). (Clearly, if we define $J_rA > 0$ and $\Delta\tilde{\mu}_i$ as positive, J_i is negative throughout this region.)

The quadrant immediately above this one, delineated by the origin, the value $(J_i/J_rZ) = 1$, and the value $(\Delta\tilde{\mu}_iZ/A) = 1$, may be referred to as the *retarding region* since in this region $J_i\,\Delta\tilde{\mu}_i > 0$ but J_i decreases with increasing q or decreasing force ratio (i.e., when A increases at constant $\Delta\tilde{\mu}_i$ or when $\Delta\tilde{\mu}_i$ decreases at constant A).

At the interface between the retarding region and the driving region $J_i = 0$. But $\Delta\tilde{\mu}_i = 0$ *only when* $q = 0$; for all nonzero values of q, $\Delta\tilde{\mu}_i > 0$. Thus, coupling to an exergonic chemical reaction establishes and maintains an electrochemical potential difference for i in the absence of a net flow of i; the greater the degree of coupling, the greater will be $\Delta\tilde{\mu}_i$ at fixed A. These points of intersection with the abscissa are referred to as *static heads.* The expenditure of metabolic energy to maintain differences in electrochemical

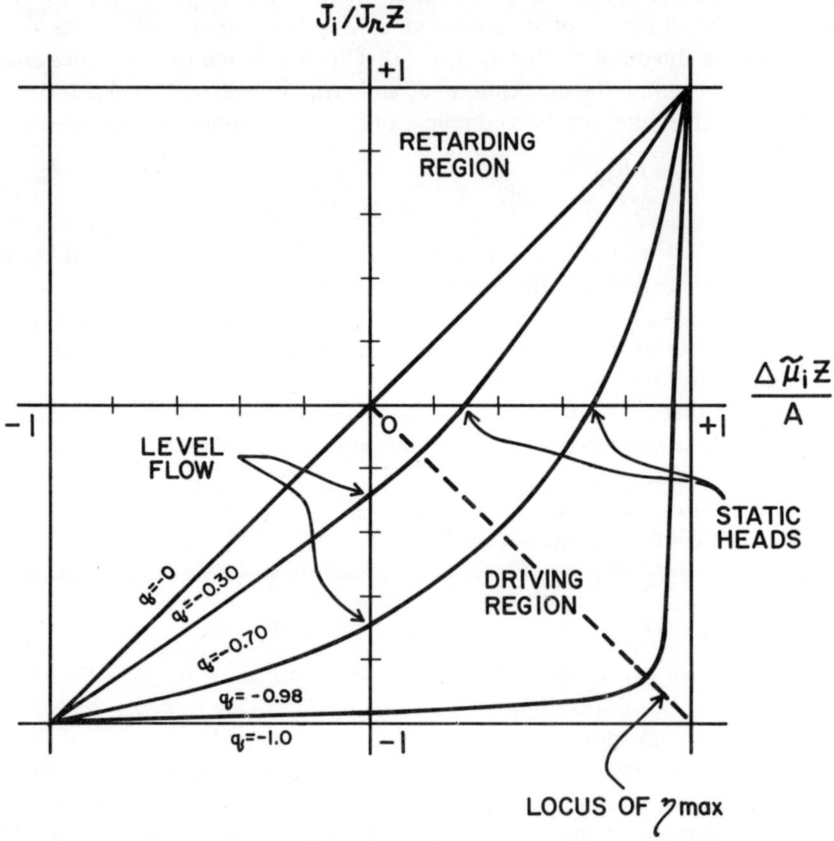

Figure 5.2. Relations between flow ratio and force ratio for different degrees of negative coupling (modified from Kedem and Caplan, 1965).

potentials in the absence of net flows is characteristic of all symmetrical cells under steady-state conditions (i.e., when cell composition is constant).

In addition, when $\Delta\tilde{\mu}_i = 0$, $J_i = 0$ only when $q = 0$; for all nonzero values of q, there is a finite flow of J_i in the direction opposite to J_r that increases with increasing q. A net flow of i when $\Delta\tilde{\mu}_i = 0$ is referred to as *level flow*, inasmuch as energy is invested to bring about a flow of matter in the absence of an electrochemical potential difference and, thus, without the performance of osmotic work. A number of epithelial tissues such as gallbladder, renal proximal tubule, and small intestine invest metabolic energy in order to absorb Na and Cl against minimal electrochemical potential differences and operate close to level flow.

Thus, the two boundaries of the driving region are the conditions, $J_i = 0$ when $\Delta\tilde{\mu}_i \neq 0$ *and* $J_i \neq 0$ when $\Delta\tilde{\mu}_i = 0$. Both are dependent upon coupling to a source of energy, but in neither instance is useful, or osmotic, work performed; thus, the efficiency of energy conversion [see equation (1.40)] is zero. Between these boundaries chemical energy is converted into osmotic work and the efficiency of the process is given by

$$\eta = -(J_i \, \Delta\tilde{\mu}_i / J_r A) \tag{5.26}$$

Substituting equations (5.24) into equations (5.22), we obtain

$$J_i = \frac{\Delta\tilde{\mu}_i + (q/Z)\,A}{R_{ii}(1 - q^2)} \tag{5.27}$$

and

$$J_r = \frac{(q/Z)\,\Delta\tilde{\mu}_i + (1/Z)^2\,A}{R_{ii}(1 - q^2)} \tag{5.28}$$

From these expressions it follows that the static head electrochemical potential difference is given by

$$(\Delta\tilde{\mu}_i)_0 = (\Delta\tilde{\mu}_i)_{J_i=0} = -qA/Z = -qA(R_{ii}/R_{rr})^{1/2} \tag{5.29}$$

and the rate of energy expenditure (power) needed to maintain this static head is

$$(J_r A)_{J_i=0} = \frac{(\Delta\tilde{\mu}_i)_0{}^2}{R_{ii}}\left(\frac{1}{q^2} - 1\right) \tag{5.30}$$

Thus, the rate of energy investment needed to maintain a given static head increases if the passive resistance to the flow of i decreases and/or the degree of coupling decreases.

The level flow is given by

$$(J_i)_{\Delta\tilde{\mu}_i=0} = \frac{-R_{ir}J_r}{R_{ii}} = \frac{A(q/Z)}{R_{ii}(1 - q^2)} \tag{5.31}$$

and the rate of energy expenditure needed to sustain this flow is

$$(J_r A)_{\Delta\tilde{\mu}_i=0} = R_{ii}(J_i)^2_{\Delta\tilde{\mu}_i=0}/q^2 \tag{5.32}$$

Thus, the rate of energy investment needed to maintain a given level flow increases if the resistance to flow increases and/or the degree of coupling decreases.

Since the efficiency of energy conversion is zero at each of the two boundaries of the driving region, it follows that η must pass through a maximum between these boundaries. It can be readily shown that this *maximum efficiency* is solely a function of the degree of coupling and is given by

$$\eta_{\max} = q^2/[1 + (1 - q^2)^{1/2}]^2 \tag{5.33}$$

Thus, $\eta_{\max} = 1$ when $q = \pm 1$ but falls very quickly with decreasing values of q. For example, when $q = 0.95$, $\eta_{\max} = 0.52$, so that extremely tight coupling would seem to be a requisite property of efficient energy conversion processes.

Finally, these relations suggest experimental approaches for determining A *if it is constant.* From equations (5.27), (5.28), and (5.29) it can be shown that

$$A = -Z(\Delta\tilde{\mu}_i)_0/q = (\partial J_i/\partial J_r)(\Delta\tilde{\mu}_i)_0 \tag{5.34}$$

Thus, determination of the relation between J_i and J_r under conditions close to static head provides a measure of A. Alternatively, from equations (5.27) and (5.28), it can be shown that

$$A = -[\partial(\Delta\tilde{\mu}_i)/\partial J_r]\ (z_i\mathfrak{F}J_i)_{\Delta\tilde{\mu}_i=0} \tag{5.35}$$

Thus, determination of the relation between J_r and $\Delta\tilde{\mu}_i$ under conditions close to level flow provides another measure of A. It should be noted that the determination of A using equations (5.34) or (5.35) involve independent measurements. Thus, obtaining the same value of A using both methods suggests that A is constant throughout the driving region and supports the internal consistancy of the model.

In concluding this section, it should be reemphasized that the model proposed by Essig and Caplan rests on several simplifying assumptions that are open to question. These investigators have summarized the body of experimental data, largely drawn from studies on isolated frog skin and toad urinary bladder, that are consistent with this linear phenomenologic approach and have derived values for A in these systems that appear to be quite reasonable (cf. Caplan and Essig, 1977). However, further study is necessary to substantiate this simple model and to establish its range of applicability. In any event the approach employed by Kedem, Caplan, and Essig provides a formal foundation for the appreciation of energy conversion resulting from the interactions of flows and forces.

6 Carrier-mediated transport

6.1. A definition of the term *carrier*

The notion that biological membranes possess components capable of interacting with a solute at one side, conveying it across the barrier and releasing it at the other side, was introduced in order to account for a number of characteristics of many biological transport processes that seemingly could not be attributed to simple diffusion. These characteristics *include* structural specificity (e.g., chemical and stereoisomeric specificity), saturation kinetics, and competitive inhibition.[1] However, as discussed in Section 2.6, diffusion through pores or channels that contain specific binding sites can also exhibit many of these characteristics.

The essential difference between a pore and a carrier, as these terms are generally employed, is that a pore has openings on both sides of the membrane so that molecules in the two surrounding solutions have *simultaneous* access to the transport pathway. In contrast, the binding site of a carrier is accessible from only one side of the membrane at a time and alternates between the two sides. Thus, a carrier necessarily involves a mobile or transitional element, whereas all of the structural properties of a pore can, in principle, be stationary or fixed. The precise nature of this "mobile" or "transitional" property is unknown; a variety of models that have been suggested are described in the excellent review by LeFevre (1975). In any event, the restriction that solutes at both sides of the membrane *do not have simultaneous access* to the binding site is the sine qua non of the carrier hypothesis that distinguishes this transport mechanism from specific pores or channels.

The kinetic behavior of carriers is generally described using schemes such as that illustrated in Figure 6.1. It is assumed that the carrier X is (a) confined to the membrane; (b) can exist in the free or bound (iX) forms; and (c) that the total amount of carrier per unit area is constant. In addition, it is frequently assumed that the association-dissociation reactions occur only at the membrane interfaces.[2] Carriers facing the outer compartment are designated X^o or iX^o and those facing the inner compartment are designated X^i or iX^i. The k_1 and k_2 are association rate constants for the formation of the complex iX; k_{-1} and k_{-2} are the corresponding dissociation rate constants. P_1 and

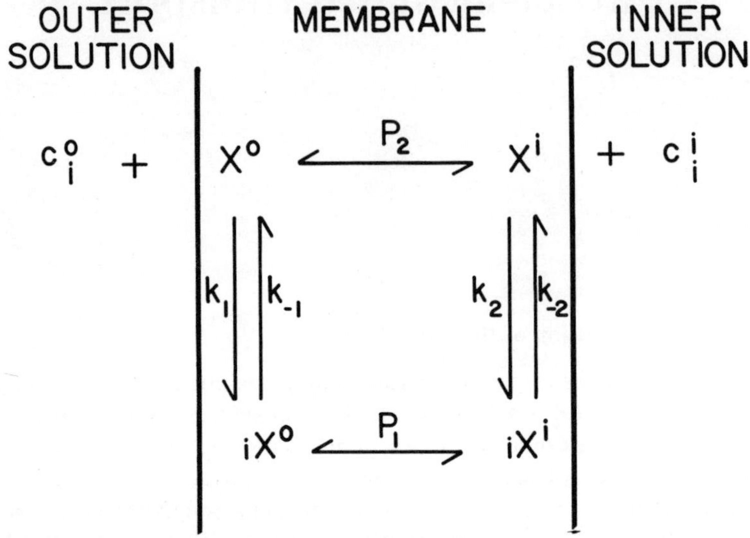

Figure 6.1. General model of carrier-mediated transport involving a carrier X that can form a binary complex with the solute i.

P_2 are the rate constants for the translocation of iX and X across the barrier; translocation is assumed to be a first-order process, but its mechanism need not be specified.

Although this model can be solved generally using the familiar approach of chemical kinetics, for the sake of simplicity we will assume that translocation through the membrane is slow compared to the association-dissociation reactions at the interfaces (i.e., the translocation step is rate limiting). Under these conditions the reactions at the interfaces are close to equilibrium and can be described by

$$K_1 = c_i{}^o [X^o] / [iX^o] \qquad (6.1)$$

and

$$K_2 = c_i{}^i [X^i] / [iX^i] \qquad (6.2)$$

where K_1 and K_2 are dissociation constants given by (k_{-1}/k_1) and (k_{-2}/k_2), respectively, and the terms in brackets are *concentrations* of free or bound carrier per unit area of membrane.

The assumption that carrier is confined to the membrane and that the amount is fixed is given by the relation

$$[X_t] = [X^o] + [iX^o] + [X^i] + [iX^i] \qquad (6.3)$$

where $[X_t]$ is the total concentration of all forms of X and is constant.

Finally, under steady-state conditions, the distribution of X at the two interfaces is constant (time independent) so that

$$P_1([iX^o] - [iX^i]) = P_2([X^i] - [X^o]) \tag{6.4}$$

(i.e., the rate of flow of bound carrier from o to i must be equal to the rate of return of free carrier from i to o).

These four independent equations permit us to solve for the concentrations of the four possible carrier forms (X^o, iX^o, X^i, and iX^i) in terms of c_i^o, c_i^i, K_1, K_2, P_1, and P_2.

The unidirectional flux of i from compartment o to compartment i is then simply $J_i^{oi} = P_1[iX^o]$; the unidirectional flux in the opposite direction is $J_i^{io} = P_1[iX^i]$; and the next flux is given by $J_i = J_i^{oi} - J_i^{io}$.

We will now consider some of the classical categories of carrier-mediated transport in terms of this kinetic approach. We will first consider the transport of uncharged solutes and later deal with charged species.

6.2. Facilitated transfer

The term *facilitated transfer* was introduced to describe carrier-mediated transport processes that are not directly coupled to nonconjugate sources of energy and, thus, cannot lead to transport against an electrochemical potential difference; they can only lead to equilibration of the transported solute across the membrane and are, therefore, sometimes referred to as *facilitated diffusion.* These processes can be described by the model illustrated in Figure 6.1 and Equations (6.1) to (6.4) when $K_1 = K_2 = K$. Under this condition it can readily be shown that

$$J_i^{oi} = \frac{P_1[X_t]c_i^o(P_2K + P_1c_i^i)}{c_i^o(P_2K + P_1K + 2P_1c_i^i) + K(2P_2K + P_2c_i^i + P_1c_i^i)} \tag{6.5}$$

and

$$J_i^{io} = \frac{P_1[X_t]c_i^i(P_2K + P_1c_i^o)}{c_i^i(P_2K + P_1K + 2P_2c_i^o) + K(2P_2K + P_1c_i^o + P_2c_i^o)} \tag{6.6}$$

Since the denominators in equations (6.5) and (6.6) are identical,

$$J_i^{oi} - J_i^{io} = J_i = \frac{[X_t](c_i^o - c_i^i)P_1P_2K}{c_i^o(P_2K + P_1K + 2P_1c_i^i) + K(2P_2K + P_2c_i^i + P_1c_i^i)} \tag{6.7}$$

Thus, when $c_i^o = c_i^i$, $J_i = 0$. As will be shown (Section 6.5), this limiting condition is entirely because $K_1 = K_2$ so that the carrier process is symmetrical.

According to equation (6.5), when c_i^i is constant, J_i^{oi} will be a saturable function of c_i^o given by

$$J_i^{oi} = J_i^m c_i^o/(K_t + c_i^o) \tag{6.8}$$

where J_i^m is the maximum velocity and is given by

$$J_i^m = \frac{P_1[X_t]\,(P_2K + P_1c_i^i)}{P_2K + P_1K + 2P_1c_i^i} \tag{6.9}$$

and K_t is the value of c_i^o at which $J_i^{oi} = J_i^m/2$, given by

$$K_t = \frac{K(2P_2K + P_2c_i^i + P_1c_i^i)}{P_2K + P_1K + 2P_1c_i^i} \tag{6.10}$$

When $P_1 = P_2 = P$, equation (6.9) reduces to $J_i^m = P[X_t]/2$; equation (6.10) reduces to $K_t = K$; and the unidirectional fluxes are given by

$$J_i^{oi} = \frac{c_i^o J_i^m}{K + c_i^o} \quad \text{and} \quad J_i^{io} = \frac{c_i^i J_i^m}{K + c_i^i} \tag{6.11}$$

Because of the similarity between equation (6.8) and the familiar Michaelis–Menten formulation of enzyme kinetics, the K_t's determined experimentally from plots of J_i^{oi} versus c_i^o or J_i^{io} versus c_i^i are frequently identified with the equilibrium constant of the interaction between the transported solute and the carrier (K). It should be clear from equation (6.10) that this is true *only when* $P_1 = P_2$; when $P_1 \neq P_2$, the experimentally determined K_t is not equal to K, but is a rather complex function of K, the P's, and either c_i^o or c_i^i.

Thus, before an experimentally determined K_t can be interpreted as a measure of the association or dissociation constant of the solute-carrier interaction, we must establish the identity between P_1 and P_2.

How can this identity be established?

It is clear from a comparison of equations (6.5) and (6.11) that when $P_1 = P_2$, J_i^{oi} is dependent on c_i^o but is not influenced by c_i^i; similarly, J_i^{io} is independent of c_i^o. In contrast, when $P_1 \neq P_2$, J_i^{oi} is influenced by c_i^i and J_i^{io} is influenced by c_i^o; these influences are referred to as *transeffects* because the unidirectional flux is affected by the concentration of solute in the (*trans*) compartment toward which it is directed as well as the (*cis*) compartment from which it originates.

When $P_1 \gg P_2$, equation (6.5) reduces to

$$J_i^{oi} \cong \frac{P_1[X_t]c_i^o}{c_i^o[2 + (K/c_i^i)] + K} \tag{6.12}$$

thus, when c_i^o is constant, J_i^{oi} increases with increasing c_i^i. This phenomenon is referred to as *transstimulation*, inasmuch as increasing the concentration in compartment i stimulates the unidirectional flux from compartment o.

Further, from equation (6.7) we see that $J_i = 0$ when $P_2 = 0$. Under this limiting condition the bound carrier iX simply shuttles back and forth across the membrane and mediates the obligatory one-for-one exchange of solute in compartment o for solute in compartment i without net transport; this process is referred to as *exchange diffusion* (see page 85).

When $P_2 \gg P_1$ equation (6.5) reduces to

$$J_i^{oi} \cong \frac{P_1 [X_t] c_i^o}{2K + c_i^o + c_i^i} \tag{6.13}$$

Thus, when c_i^o is constant, J_i^{oi} decreases with increasing c_i^i. This phenomenon is referred to as *transinhibition* and occurs when the translocation of bound carrier is rate limiting so that an increase in c_i^i leading to an increase in iX^i reduces the amount of free carrier available for recycling across the membrane.

Thus, in principle, experiments can be designed to determine whether an equilibrating system exhibits transeffects and the absence of such effects must be demonstrated before we can consider K_t a measure of the "affinity" of the carrier for the solute.

6.3. Competitive interactions and countertransport

Consider the system illustrated in Figure 6.2, where a carrier is capable of interacting with two uncharged solutes i and j, with K_1 being the dissociation constant of iX and K_2 the dissociation constant of jX. The six equations that describe this system are

$$\begin{aligned} c_i^o [X^o] - K_1 [iX^o] &= 0 \\ c_j^o [X^o] - K_2 [jX^o] &= 0 \\ c_i^i [X^i] - K_1 [iX^i] &= 0 \\ c_j^i [X^i] - K_2 [jX^i] &= 0 \end{aligned} \tag{6.14}$$

$$[X^o] + [X^i] + [iX^o] + [iX^i] + [jX^o] + [jX^i] = [X_t]$$

and

$$P_1 [iX^o] - P_1 [iX^i] + P_2 [X^o] - P_2 [X^i] + P_3 [jX^o] - P_3 [jX^i] = 0$$

With some patience, these six equations can be solved for the six possible carrier forms in terms of P's, K's and concentrations, and expressions for the unidirectional fluxes of i and j can be derived as above.

It can be readily shown that when $P_1 = P_2 = P_3 = P$

$$J_i^{oi} = \frac{J_i^m c_i^o}{K_1 [1 + (c_j^o/K_2)] + c_i^o} \tag{6.15}$$

which is precisely analogous to the expression for classical *competitive inhibition* in enzyme kinetics. Clearly, J_i^{oi} decreases with increasing c_j^o and

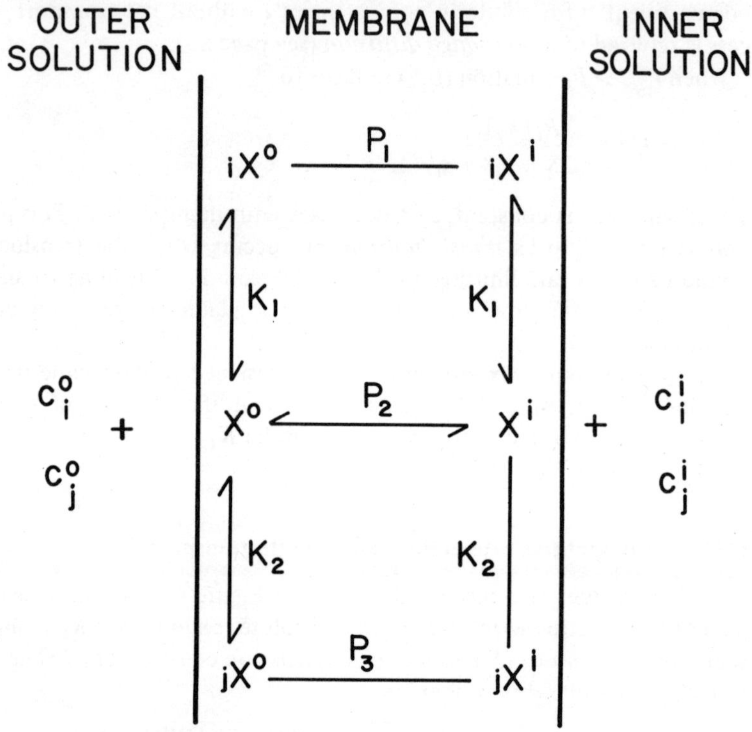

Figure 6.2. Model of competitive inhibition in a system capable only of facilitated transfer.

when $c_j{}^o = K_2$ the concentration of i needed to achieve a half-maximal velocity is $2K_1$ (i.e., it has been doubled).

Let us now consider the case in which $c_j{}^o = 0$ but $c_j{}^i > 0$. From equation (6.15) it follows that

$$J_i{}^{oi} = \frac{J_i{}^m c_i{}^o}{K_1 + c_i{}^o} \tag{6.16}$$

and

$$J_i{}^{io} = \frac{J_i{}^m c_i{}^i}{K_1[1 + (c_j{}^i/K_2)] + c_i{}^i} \tag{6.17}$$

Thus,

$$J_i{}^{oi} - J_i{}^{io} = J_i = J_i{}^m \left\{ \frac{c_i{}^o}{K_1 + c_i{}^o} - \frac{c_i{}^i}{K_1[1 + (c_j{}^i/K_2)] + c_i{}^i} \right\} \tag{6.18}$$

We now inquire: When will J_i cease? By setting equation (6.18) equal to zero and rearranging, it can be shown that $J_i = 0$ when

$$(c_i^i/c_i^o) = 1 + (c_j^i/K_2) \tag{6.19}$$

Thus, in spite of the fact that this is an equilibrating system, it is capable of transporting i from compartment o to compartment i against a concentration difference as long as $c_j^i > 0$. In general, when $c_j^o \neq 0, J_i$ ceases when

$$(c_i^i/c_i^o) = (K_2 + c_j^i)/(K_2 + c_j^o) \tag{6.20}$$

so that the system is capable of transporting i against a chemical potential difference, providing $c_j^i > c_j^o$.

This phenomenon, which was described formally by Wilbrandt and Rosenberg (1957, 1961) is referred to as *countertransport.*

The mechanism underlying countertransport is intuitively obvious! In the simple case where $c_j^o = 0, J_i^{oi}$ is unimpeded, whereas J_i^{io} is retarded by competition between i and j in compartment i for free carrier. This emerges clearly by substituting the relation $K_2 = c_j^i[X^i]/[jX^i]$ into equation (6.19), which gives

$$(c_i^i/c_i^o) = 1 + ([jX^i]/[X^i])$$

Since the total amount of carrier is fixed, as the ratio of $[jX^i]$ to $[X^i]$ increases, the amount of free carrier available to mediate the movement of i from compartment i to compartment o decreases so that i becomes "trapped" in compartment i. Thus, when $c_j^i > c_j^o$, J_i^{oi} will equal J_i^{io} only when $c_i^i > c_i^o$.

However, it should be clear from our previous considerations that coupling to a nonconjugate source of energy is required for any mechanism to transport matter against an electrochemical potential difference; what is the source of this requisite energy input in the case of countertransport? This issue will be dealt with more completely later (Section 6.4); suffice it to say at this point that the energy for the flow of i against its concentration difference is derived from coupling to the flow of j from a higher to a lower concentration where the coupling mechanism is the common carrier system. In short, because j and i share a common carrier, the downhill flow of j from compartment i to compartment o can provide the energy for the uphill flow of i in the opposite direction.

This phenomenon can be illustrated more vividly by considering the condition where $P_2 = 0$. Clearly, then, in the absence of j, $J_i = 0$ and the system is only capable of mediating exchange diffusion. When j is present in one or both solutions, in general, there will be a net flow of i, which under steady-state conditions must equal the net flow of j in the opposite direction; in short, this system will carry out an obligatory one-for-one exchange of i and

j. When $P_1 = P_3 = P$ and $P_2 = 0$, the solutions of equations (6.14) are

$$J_i^{oi} = P[X_t]\ K_2 c_i^o (K_2 c_i^i + K_1 c_j^i)/D \qquad (6.21)$$

and

$$J_i^{io} = P[X_t]\ K_2 c_i^i (K_2 c_i^o + K_1 c_j^o)/D \qquad (6.22)$$

where

$$D = 2K_1 K_2 c_j^o c_i^i + 2K_1^2 c_j^o c_j^i + K_1 K_2 c_i^o c_j^i + K_1 K_2^2 (c_i^i + c_i^o) + K_1^2 K_2 (c_j^o + c_j^i) + 2K_2^2 c_i^o c_i^i \qquad (6.23)$$

J_j^{oi} and J_j^{io} are given by equations (6.21) and (6.22) after replacing c_i^o by c_j^o, c_i^i by c_j^i, and K_1 by K_2; the denominator remains the same.

It follows from equations (6.21) and (6.22) that $J_i = 0$ when

$$(c_i^i/c_i^o) = (c_j^i/c_j^o) \qquad (6.24)$$

and under these conditions J_j is also zero. This is the static head state for an *obligatory*, one-for-one, countertransport system. Since $\Delta\mu_i = RT\ln(c_i^i/c_i^o)$ and $\Delta\mu_j = RT\ln(c_j^i/c_j^o)$, it is obvious that the condition of zero flows is reached when $\Delta\mu_i = \Delta\mu_j$. If $(c_j^i/c_j^o) > (c_i^i/c_i^o)$, there will be a flow of i from compartment o to compartment i against a concentration difference *driven* by an equal flow of j in the opposite direction from a higher to a lower concentration. Conversely, if $(c_i^i/c_i^o) > (c_j^i/c_j^o)$, there will be a flow of j from o to i against a concentration difference driven by an oppositely directed flow of i.

It is instructive to compare equation (6.20), which applies when $P_2 \neq 0$, with equation (6.24), which applies when $P_2 = 0$. Clearly, for any finite value of K_2 when $c_j^i > c_j^o$

$$(K_2 + c_j^i)/(K_2 + c_j^o) < (c_j^i/c_j^o)$$

Thus, when $P_2 \neq 0$, the static head value of $\Delta\mu_i$ will be less than $\Delta\mu_j$. That is, because X can cross the barrier in the free form, the flow of i is not *tightly coupled* to the flow of j and the efficiency of energy conversion is less than the ideal of unity. Thus, a high efficiency of energy conversion in a countertransport system requires that the free carrier be immobile.

Finally, there is an obvious point that, nonetheless, should be emphasized. In the "real world" where the efficiency of energy conversion is less than unity, some input of metabolic energy is necessary in order to maintain steady-state differences in chemical potential; the countertransport system illustrated in Figure 6.2 cannot accomplish this by itself! For example, let us suppose that at the outset of an experiment, $c_i^o = c_i^i$ and $c_j^i > c_j^o$. Then, there will be a flow of i from compartment o to compartment i so that we will achieve the condition $(c_i^i/c_i^o) > 1$. However, in the real world the condition $c_j^i > c_j^o$ cannot be sustained indefinitely unless some other mechanism intervenes; otherwise, this concentration difference will ultimately vanish, however slowly. Consequently, the final steady-state condition of this isolated system is $(c_i^i/c_i^o) = (c_j^i/c_j^o) = 1$. The only way a true steady state can be sustained, where (c_j^i/c_j^o) and (c_i^i/c_i^o) are greater than unity, is if there is some other mechanism capable of replenishing j in compartment i. For example, j may be the product of a chemical reaction in compartment i or there could be a mechanism capable of actively transporting j from compartment o to compartment i. In any event, suffice it to say that a constant supply of energy is necessary to maintain any system in a steady state that is displaced from equilibrium.

6.4. Cotransport

Countertransport (sometimes referred to as "antiport") is one mechanism for carrier-mediated coupling between flows of solutes that can lead to transport against a concentration gradient. Another mechanism that is prevalent in biological systems (cf. Schultz and Curran, 1970) is cotransport (or "symport") where the net movements of two solutes take place in the same direction. Consider the general scheme illustrated in Figure 6.3, where i and j can combine with the same carrier X to form the binary complexes iX and jX and the terniary complex ijX. In the most general case, the order of binding is random and all forms of the carrier can translocate across the barrier. We can, of course, impose a *preferred order* of binding by adjusting the values of the K's and can restrict the carrier forms capable of translocation by setting one or more of the P's equal to zero. The resulting expressions for the fluxes of i and j can become cumbersome and their derivations are rather tedious, but the underlying approach is the same as that outlined previously for simpler systems. The general solutions of the model for a number of selected conditions are given by Schultz and Curran (1970).

In order to illustrate the principal features of cotransport, let us consider two simple examples.

Example I. Assume that i must bind to X first to form the binary complex iX and that only then can j be bound to form ijX and that the terniary complex dissociates in the reverse order (i.e., j leaves first); this restriction is achieved by setting the dissociation constant $K_4 = \infty$ so that $jX^o = 0$. The system then reduces to that enclosed by the dashed rectangle in Figure 6.3. Let us further assume that $P_1 = P_2 = P_3 = P$.

The solution of this model for the bidirectional fluxes of i and j are

$$J_i^{oi} = \frac{J_i^m c_i^o}{[K_1 K_2/(K_2 + c_j^o)] + c_i^o} \tag{6.25}$$

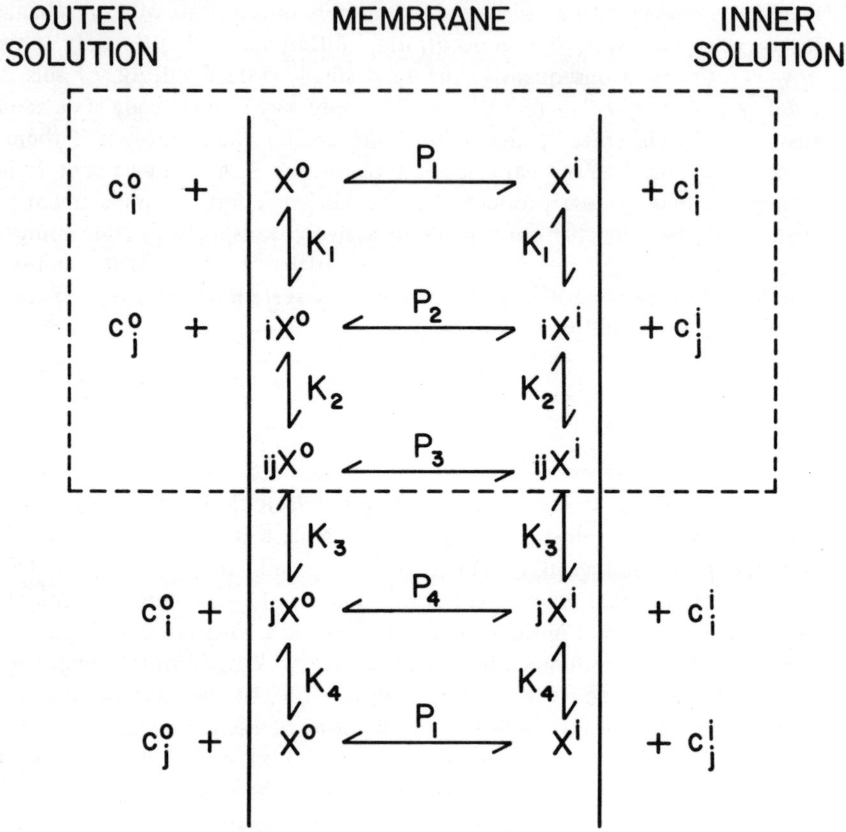

Figure 6.3. General model for cotransport of the solutes *i* and *j*.

$$J_i^{io} = \frac{J_i^m c_i^i}{[K_1 K_2/(K_2 + c_j^i)] + c_i^i} \tag{6.26}$$

$$J_j^{oi} = \frac{J_i^m c_i^o c_j^o}{K_1 K_2 + c_i^o(K_2 + c_j^o)} = \frac{J_i^{oi} c_j^o}{K_2 + c_j^o} \tag{6.27}$$

and

$$J_j^{io} = \frac{J_i^m c_i^i c_j^i}{K_1 K_2 + c_i^i(K_2 + c_j^i)} = \frac{J_i^{io} c_j^i}{K_2 + c_j^i} \tag{6.28}$$

where $J_i^m = [X]_t P$. (It should be noted that there are no transeffects because the P's are equal.)

Thus, the maximum velocity of i in either direction is independent of j but the concentration needed to achieve a half-maximal velocity is a function of j;

clearly, when $c_j = 0$, $K_t = K_1$ as is to be expected. On the other hand, the maximum velocity of j in either direction is a direct function of the concentration of i in the cis solution. Further, from equations (6.27) *and* (6.28) we see that

$$\frac{J_j^{oi}}{J_i^{oi}} = \frac{c_j^o}{(K_2 + c_j^o)} \quad \text{and} \quad \frac{J_j^{io}}{J_i^{io}} = \frac{c_j^i}{(K_2 + c_j^i)} \tag{6.29}$$

The expressions on the right-hand sides of these equations are referred to as the *coupling coefficients* between the flows of i and j.

It follows from equations (6.25) and (6.26) that $J_i = 0$ when $(c_i^i/c_i^o) = (K_2 + c_j^o)/(K_2 + c_j^i)$; the resemblance between this expression and equation (6.20) for the countertransport system should be noted. Thus, $(c_i^i/c_i^o) > 1$ when $c_j^o > c_j^i$; that is, the system is capable of transporting i against a concentration difference and the required energy is derived from carrier-mediated coupling to the flow of j in the same direction from a higher to a lower concentration.

From equations (6.27) and (6.28) it can be shown that $J_j = 0$ when

$$\frac{c_j^i}{c_j^o} = \frac{c_i^o(K_1 + c_i^i)}{c_i^i(K_1 + c_i^o)} \tag{6.30}$$

Thus, $(c_j^i/c_j^o) > 1$ when $c_i^o > c_i^i$, so that j can be transported against a concentration difference energized by coupling to the flow of i down a concentration difference.

Example II. Now let us consider the same sequence of binding and dissociation but with the restriction that $P_2 = 0$ so only the free carrier and the terniary complex can translocate. Under these conditions the bidirectional fluxes of i and j are complicated by transconcentration effects; the explicit solutions can be found in the review by Schultz and Curran (1970). The important point is that under these conditions the *coupling coefficient* is unity so that $J_i^{oi} = J_j^{oi}$, $J_i^{io} = J_j^{io}$ and $J_i = J_j$. Further, J_i and J_j will both cease when $(c_i^i/c_i^o) = (c_j^o/c_j^i)$; again, this expression should be compared with equation (6.24) for the analogous countertransport system.

Since for any finite value of K_2 when $c_j^o > c_j^i$

$$[(K_2 + c_j^o)/(K_2 + c_j^i)] < (c_j^o/c_j^i)$$

it follows that the static head value of (c_i^i/c_i^o) will be less when $P_2 \neq 0$ than when $P_2 = 0$.

Once again, when $P_2 = 0$, the flows of i and j are tightly coupled and the energy derived from the flow of one of these species from a higher to a lower concentration can be completely utilized to drive the flow of the other species from a lower to a higher concentration. When $P_2 \neq 0$, the flows are not tightly coupled and the efficiency of energy conversion is reduced.

In short, the principles that underlie carrier-mediated cotransport processes are strictly analogous to those that govern the behavior of countertransport processes. In both instances the energy derived from the flow of one species down an electrochemical potential difference can be employed to propel the flow of the other species against an electrochemical potential difference; the coupling mechanism responsible for this energy conversion (or transduction) is a *common* carrier system, and the efficiency of energy conversion is, as discussed in Section 5.3, determined by the tightness (or degree) of coupling between the flows.[3]

6.5. "Primary" active transport

All of the carrier models we have discussed to this point are inherently symmetrical. We have seen that in the cases of cotransport and countertransport, a solute i can be propelled from a lower to a higher concentration when there is an asymmetric distribution of the "coupled" species j. We referred to these processes as *secondary active transport* because of solute-solute (or osmo-osmotic) coupling. In the case of primary active transport the carrier system is inherently asymmetric and this essential asymmetry arises from coupling the system to an exergonic chemical (metabolic) reaction. In order to illustrate this point, let us consider the hypothetical model shown in Figure 6.4.[4] We assume that the total amount of free carrier can exist in either of two mobile forms; a form designated by X which (as before) can combine with the solute i, and a form designated by Y which is incapable of binding i. We further assume that the distributions of total free carrier between X and Y at each interface are given by the equilibrium constants $K_2 = ([Y^o]/[X^o])$ and $K_3 = ([Y^i]/[X^i])$. The solution of the equations that describe this system for J_i is simply

$$J_i = J_i^{oi} - J_i^{io} = J_i^m \left\{ \left[\frac{c_i^o}{K_1(1+K_2) + c_i^o} \right] - \left[\frac{c_i^i}{K_1(1+K_3) + c_i^i} \right] \right\} \tag{6.31}$$

Thus, both unidirectional fluxes conform to Michaelis-Menten kinetics, where the K_t for J_i^{oi} is $K_1(1 + K_2)$ and that for J_i^{io} is $K_1(1 + K_3)$. Clearly, if $K_2 \neq K_3$ the system is inherently asymmetric, inasmuch as the concentration of i in the outer solution needed to achieve a half-maximal flow from compartment o to compartment i will differ from the concentration of i in the inner solution needed to achieve a half-maximal flow in the opposite direction.

From equation (6.31) we see that when $J_i = 0$

$$(c_i^i/c_i^o) = (1 + K_3)/(1 + K_2) \tag{6.32}$$

Thus, when $K_3 > K_2$, the system is capable of transporting i from a lower concentration in compartment o to a higher concentration in compartment i; if $K_3 < K_2$, i will be *actively* transported in the opposite direction.

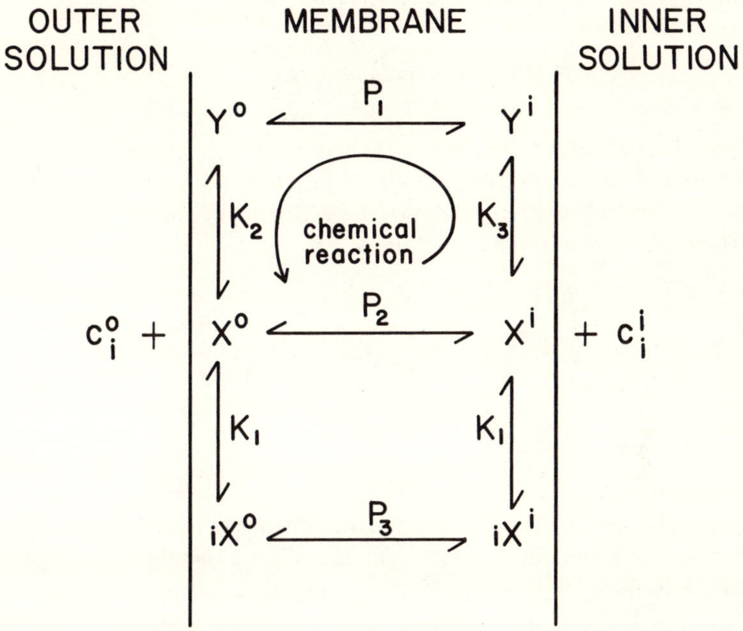

Figure 6.4. Highly simplified model of a system capable of active transport.

Although this conclusion follows directly from kinetic considerations, what about the energetics? Some insight into the energetics of this simple, primary active transport system can be gained by describing the static head condition using the formalism of irreversible thermodynamics.

In general, the dissipation function for this system is given by

$$T\,d_iS/dt = J_{iX}X_{iX} + J_XX_X + J_YX_Y$$

where, as before, J's are flows, the X's are the conjugate forces and the subscripts designate the species. When $J_i = 0$, the static head condition, $J_{iX} = 0$ so that this expression reduces to

$$\frac{T\,d_iS}{dt} = P([X^o] - [X^i])RT\ln\left(\frac{[X^o]}{[X^i]}\right) + P([Y^o] - [Y^i])RT\ln\left(\frac{[Y^o]}{[Y^i]}\right) \tag{6.33}$$

Under steady-state conditions when $J_{iX} = 0$,

$$P([X^o] - [X^i]) = -P([Y^o] - [Y^i])$$

so that equation (6.33) reduces to

$$\frac{T\,d_iS}{dt} = -P([Y^o] - [Y^i])RT\ln\left(\frac{[X^o]\,[Y^i]}{[X^i]\,[Y^o]}\right) \tag{6.34}$$

Since $[Y^o] = K_2[X^o]$ and $[Y^i] = K_3[X^i]$, equation (6.34) may be written

$$\frac{T\,d_iS}{dt} = -P([Y^o] - [Y^i])RT \ln\left(\frac{K_3}{K_2}\right) \tag{6.35}$$

Now let us assume, as illustrated in Figure 6.4, that a chemical reaction is responsible for converting X^i to Y^i and that Y^i then translocates from the inner interface to the outer interface where it spontaneously re-forms X^o. Then, under steady-state conditions

$$-J_r = P([Y^o] - [Y^i])$$

where J_r (as before) is the rate of the chemical reaction. Further, it should be clear that the expression

$$RT \ln\left(\frac{[X^o]\,[Y^i]}{[X^i]\,[Y^o]}\right)$$

is simply the difference between the chemical potentials of the products, X^o and Y^i, and the reactants, X^i and Y^o, of the chemical reactions at the two interfaces or the change in free energy of the system per unit rate of reaction. Thus, we may write

$$T\,d_iS/dt = J_rA$$

where A, the affinity, is given by the sum of the chemical potentials (gradients) of the two carrier forms.

In short, when $K_3 \neq K_2$, a static head will be achieved where $c_i^o \neq c_i^i$ and free energy must be dissipated by the system to maintain this asymmetry. The energy is invested to maintain gradients of free carrier and the rate of energy expenditure is determined by the magnitudes of these gradients and the rate at which they tend to be dissipated by recycling of X and Y.

Finally, it should be noted that the maximum osmotic work that can be performed by this system is given by

$$RT \ln\left(\frac{c_i^i}{c_i^o}\right) = RT \ln\left(\frac{1 + K_3}{1 + K_2}\right)$$

and when $K_3 > K_2$ this is less than the free energy of the driving reaction (A), which is $RT\ln(K_3/K_2)$. The reason for this is that $P_2 \neq 0$, so that gradients of X established by the chemical reaction can be dissipated by the movement of X through the barrier. If $P_2 = 0$, the system is tightly (ideally) coupled and the static head that will be achieved is given by $(c_i^i/c_i^o) = (K_3/K_2)$. At this point all flows will cease inasmuch as $\Delta\mu_i$ is now equal and opposite to A. If we now perturb the system so that $(c_i^i/c_i^o) > (K_3/K_2)$, the chemical reaction may be driven in the reverse direction (Mitchell, 1970).

6.6. Rheogenic carrier-mediated transport

Up to this point we have limited our considerations to carrier-mediated transport of uncharged solutes. We will now briefly consider the cases illustrated in Figure 6.5, where a monovalent cation i is transported across a membrane after forming either (a) a positively charged complex (iX_+) with a neutral carrier (X) or (b) a neutral complex (iX) with a monovalent anionic carrier (X_-). We will use these examples to illustrate an approach toward the formal description of such systems as well as several points bearing on the distinction between carrier-mediated versus diffusional transport of charged species. Since these mechanisms result in a carrier-mediated transfer of charge, we refer to them as "rheogenic" or "current generating" (from the Greek, *rheo* = current or flow).

We assume that the presence of an electrical potential difference across the membrane only affects the translocation rate of the charged carrier form(s) and has no effect on the association-dissociation reactions at the interfaces. Thus, the equations that describe the association-dissociation reactions in these systems are

$$K[iX^o] - c_i^o[X^o] = 0$$

$$K[iX^i] - c_i^i[X^i] = 0$$

and the conservation of carrier is given by

$$[X^o] + [iX^o] + [iX^i] + [X^i] = [X_t]$$

where $[X_t]$ is constant; these equations are identical to those that describe noncharged carrier systems.

If the first-order translocation rate constant of the neutral carrier forms is given by P, then the rate constant for translocation of the charged carrier forms must be modified by the factor ξ^α or $\xi^{(\alpha-1)}$ where $\xi = \exp(\mathcal{F}\Delta\psi/RT)$. The value of α must lie between 0 and 1 and is a function of the electrical profile with the membrane. If the electrical potential barrier is symmetrical about the midpoint of the membrane, the rate constants for translocation from o to i and i to o will be equally and oppositely affected so that $\alpha = 0.5$ (see Geck and Heinz, 1976; Stein, 1977; Hall et al., 1973). Under these conditions:

(a) When the carrier is neutral, the steady state is described by

$$P\{[X^o] - [X^i] + \xi^{-1/2}[iX_+^o] - \xi^{1/2}[iX_+^i]\} = 0$$

and the unidirectional flux of i from compartment o to compartment i is given by $P\xi^{-1/2}[iX_+^o]$; the unidirectional flux of i in the opposite direction is given by $P\xi^{1/2}[iX_+^i]$.

(b) If the carrier is a monovalent anion so that the complex iX is neutral,

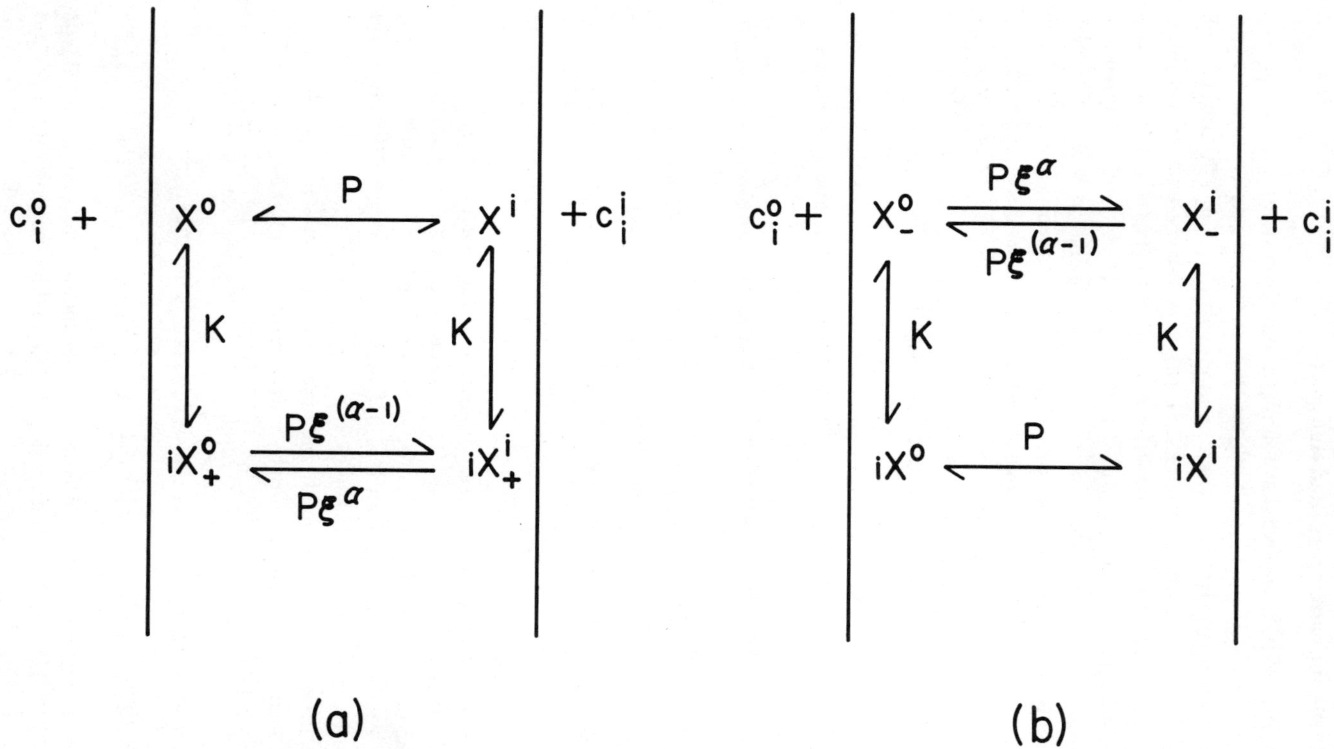

Figure 6.5. Models of carrier-mediated "rheogenic" transport. (a) Uncharged carrier X forming a charged complex iX_+ with a cation i. (b) Charged carrier X_- forming a neutral complex with a cation i.

the steady state is described by

$$P\{\xi^{1/2}[X_-^o] - \xi^{-1/2}[X_-^i] + [iX^o] - [iX^i]\} = 0$$

Solving these equations, we obtain:

(a) When the free carrier is neutral,

$$J_i^{oi} = \{P[X_t]\ c_i^o(K\xi^{-1/2} + c_i^i)\}/D \tag{6.36}$$

$$J_i^{io} = \{P[X_t]\ c_i^i(K\xi^{1/2} + c_i^o)\}/D \tag{6.37}$$

and

$$J_i = \{P[X_t]\ K(c_i^o\xi^{-1/2} - c_i^i\xi^{1/2})\}/D \tag{6.38}$$

where

$$D = 2K^2 + Kc_i^o(1 + \xi^{-1/2}) + Kc_i^i(1 + \xi^{1/2}) + c_i^o c_i^i(\xi^{1/2} + \xi^{-1/2}) \tag{6.39}$$

(b) When the free carrier is a monovalent anion J_i^{oi}, J_i^{io}, and J_i are also given by equations (6.36) to (6.38), but D is given by

$$D = 2c_i^o c_i^i + Kc_i^o(1 + \xi^{-1/2}) + Kc_i^i(1 + \xi^{1/2}) + K^2(\xi^{1/2} + \xi^{-1/2}) \tag{6.40}$$

Several points should be noted:

First, in both cases $J_i = 0$ when $(c_i^o/c_i^i) = \exp(\mathcal{F}\Delta\psi/RT)$, that is, the Nernst equilibrium distribution, as must be the case for any equilibrating system.

Second, when $\Delta\psi = 0$, equations (6.36) to (6.38) reduce to equations (6.11) for facilitated transfer of an uncharged solute.

Third, when $c_i^o = c_i^i$, the relation between J_i and $\Delta\psi$ is described by an antisymmetric curve about the origin.

Finally, in Figure 6.6 the relations between J_i and $\Delta\psi$ calculated from equations (6.38) to (6.40) are compared with that predicted by the "constant field" flux equation (2.13). In the example shown in Figure 6.6, c_i^o and c_i^i were fixed at 100 mM and 10 mM, respectively; P was set equal to 0.01 cm/h (as in Figure 2.3); K was set equal to 1 mM and X_t was adjusted so that the values calculated from equations (6.38) to (6.40) when $\Delta\psi = 0$ were equal to the value given by equation (2.13) when $\Delta\psi = 0$. Clearly, over the range $\Delta\psi = \pm 100$ mV the three curves are quite distinctive, but over the range $\Delta\psi = 0$ to $\Delta\psi = 58$ mV (the equilibrium or reversal potential) they are indistinguishable. Further, it can be readily shown that as c_i^i approaches c_i^o these curves become increasingly linear, the intersections approach the origin, and one can choose parameters that expand the range over which they are essentially indistinguishable.

There are important practical messages to be gained from this exercise. As will be discussed further (Section 7.1), experimentally determined relations between J_i (or I_i) and $\Delta\psi$ (so-called current-voltage or I–V relations) can provide a considerable amount of information regarding the properties of a

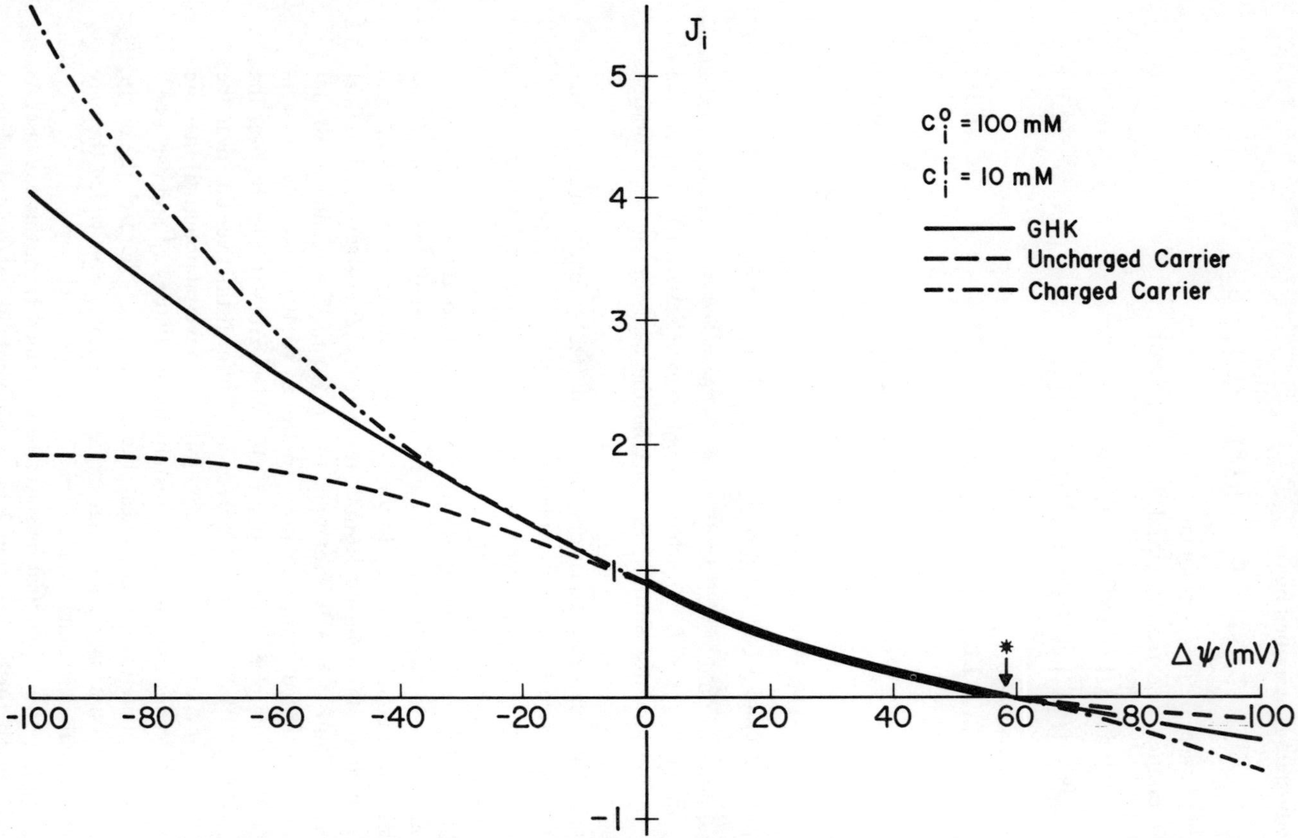

Figure 6.6. Relations between J_i and $\Delta\psi$ predicted by the GHK equation for simple diffusion of a monovalent cation and by equations (6.38) to (6.40) for carrier-mediated transport of a monovalent cation by an uncharged or charged carrier when $\alpha = 0.5$. The arrow indicates the "reversal potential" (58 mV) when the concentration ratio is 10.

system and the underlying transport mechanism. Thus, if only c_i^o is known and there is reason to believe that J_i is not influenced by nonconjugate driving forces, the value of $\Delta\psi$ when $J_i = 0$ (the reversal potential) permits us to calculate c_i^i from the Nernst relation; on the other hand, if both c_i^o and c_i^i are known and $(\Delta\psi)_{J_i=0}$ is not equal to $(RT/\mathcal{F})\ln(c_i^o/c_i^i)$, the transport process must be influenced by a nonconjugate driving force. This is true regardless of the underlying transport mechanism!

Further, as shown in Figure 6.6, the relations between J_i and $\Delta\psi$ predicted by the constant-field equation and by the two rheogenic carrier mechanisms are quite distinctive *providing studies are carried out over a sufficiently wide range of* $\Delta\psi$*'s*. As discussed in Section 2.3, if J_i is given by the constant-field equation, the value of J_i when $\Delta\psi = 0$ is a measure of P_i so that knowing $(\Delta\psi)_{J_i=0}$ and $(J_i)_{\Delta\psi=0}$ (i.e., c_i^o, c_i^i, and P_i) the entire curve can be constructed. Significant departures from this predicted curve may suggest the involvement of a neutral ion carrier or a charged carrier. However, it must be reemphasized that the relations between J_i and $\Delta\psi$ predicted by these three quite different transport mechanisms may be indistinguishable over a wide range, so that they must be interpreted cautiously.[5]

6.7. Distinction between carriers and pores

By this point, it should be clear that pore-mediated transport and carrier-mediated transport may display a number of common features, such as saturation kinetics, competitive inhibition, and transeffects. Further, as discussed in Section 6.6, the relations between J_i and $\Delta\psi$ for charged solutes may not provide clear grounds for distinguishing between simple diffusion and carrier-mediated transport. How, then, can one distinguish between these two different transport mechanisms?

Considerable insight into this problem has emerged from studies of artificial lipid bilayers "doped" with antibiotics that are capable of rendering these barriers permeable to ions; in the absence of these agents, lipid bilayers posses extremely high electrical resistances ($>10^8\ \Omega \cdot \text{cm}^2$) indicating that they are essentially impermeable to hydrophilic charged species. Two groups of such agents have been identified: (a) those that serve as mobile carriers or "ionophores" (true "ferry boats" in the classical sense) such as nonactin, valynomycin, and monensin; and (b) those that form water-filled pores that span the entire thickness of the barrier such as gramacidin *A* and amphotericin *B*. Much of the experimental evidence that ion transport through pores can exhibit saturation kinetics, competitive inhibition and transeffects is derived from studies on artificial lipid bilayers treated with gramacidin *A* (cf. Eisenman, 1973; Urban et al., 1978; Hladky, 1974).

Studies on nonactin-doped lipid membranes indicate that each carrier molecule can transport, at most, about 10^4 ions per second across a bilayer approximately 100 Å thick (Läuger, 1972). On the other hand, studies on lipid membranes doped with pore-forming agents (as well as theoretical considerations) indicate that each pore can conduct as many as 10^6 to 10^7 ions per second, a value two to three orders of magnitude greater than the maximum "turnover" number of *known* carriers (Läuger, 1972; Hladky, 1974).

If these arguments can be extrapolated to biological membranes, measurements of *single-site* conductance could distinguish between pores and mobile carriers. Estimates of the conductances of single Na entry sites in a variety of excitable membranes indicate that each site is capable of transporting approximately 10^6 ions/sec, suggesting that passive Na entry into these cells is mediated by pores or channels rather than carriers (Armstrong, 1975).

However, it must be emphasized that the notion that the turnover number of carriers is low compared to that of pores is based on experiments with artificial lipid membranes, approximately 100 Å thick, doped with ionophores whose diameters are approximately 10 to 15 Å (cf. Simon and Morf, 1973). Thus, these carriers, in fact, are small ferries that must diffuse a distance at least seven times their size through a viscous milieu in the course of shuttling one ion across the barrier and then must return for the next passenger; this is a time-consuming voyage and any lateral movements in the plane of the membrane (side trips) will increase its duration. It is reasonable to question whether these artificial carrier systems are realistic models of purported carriers in biological membranes. This question cannot be answered at this time and, in the absence of evidence to the contrary, we can certainly conceive of carrier models that do not involve the shuttling of a binding site back and forth across the entire thickness of the membrane. For example, a pore with a recognition site that *alternately* faces one side of the membrane and then the other would, according to our definition, be considered a "carrier" and might exhibit turnover times that equal or rival those observed in "true" pores. Thus, the use of "turnover number" to distinguish between pores and carriers in biological membranes may not be generally valid.

As already discussed, the essential property of carriers, as they are generally conceptualized, is that their recognition or binding sites are capable of undergoing transitions that shift their accessibility from one side of the membrane to the other. In contrast, we view pores as being simultaneously accessible to solutes from both sides. Criteria for distinguishing between pores and carriers should be *direct consequences of this fundamental difference.* In this sense, the phenomenon of "uphill" counter-transport is the most widely accepted positive criterion for the implication of a carrier rather than a pore (Widdas, 1952; Rosenberg and Wilbrandt, 1957; Wilbrandt and Rosenberg, 1961; LeFevre, 1974). It is difficult to see how, in general, the downhill flow of a solute through a pore in one direction can energize the flow of another solute in the opposite direction; yet, this phenomenon has been widely observed experimentally (LeFevre, 1974) and can be readily accounted for using simple carrier models.

In addition, it is difficult to see how some well-documented examples of cotransport, where the downhill flow of one solute energizes the uphill flow of another in the same direction, can be accommodated by a pore model. Certainly, transport through pores can exhibit positive coupling. An excellent example of this phenomenon is solvent drag (Andersen and Ussing, 1957). In addition, there are several examples of uphill cotransport that appear to be attributable to solute-solute interactions (solute drag) (Ussing, 1966; Franz and Van Bruggen, 1967; Biber and Curran, 1968), which are presumably the result of frictional interactions (momentum transfer) that take place in water-filled channels and do not involve carriers. However, in all of these instances of cotransport caused by purely mechanical interactions, the stoichiometry or coupling coefficient is highly variable. On the other hand, there are a number of well-documented cotransport processes where the coupling coefficients appear to be fixed and independent of the concentrations of either of the cotransported species or their respective conjugate driving forces (Schultz and Curran, 1970). These phenomena can be readily described using carrier models that are simple extensions of well-established enzyme reactions; it is far more difficult to see how they can be the results of transport through pores.[6]

6.8. Summary

The carrier hypothesis has led a precarious existence since its inception in 1933-35 (Osterhout, 1933, 1935; Lundegårdh, 1935); nonetheless, it has survived and now appears to be flourishing. There is a conglomerate of kinetic evidence consistent with this notion (heroically summarized by LeFevre, 1974), and there are no equally acceptable alternatives for many of these observations. During the past fifteen years this area has embarked on a new direction aimed at isolating purported carriers and characterizing their behavior after reconstitution in artificial lipid membranes. It is hoped that as a result of these efforts, abstractions that were created to satisfy kinetic observations will acquire indisputable biochemical identities.

In this chapter we introduced the types of transport phenomena generally attributed to carriers using a kinetic approach. In some instances these formalisms are qualitatively correct but do not hold up quantitatively under rigorous testing; such inconsistencies, or "anomalies," and possible explanations are summarized by LeFevre (1974). Suffice it to say, that the present stage of development of this area resembles that of enzymology before enzymes were finally isolated, purified, and characterized with respect to structure and mechanism of action; many of these problems will not be resolved until this area of membranology achieves the current state of the art of enzymology.

7 Some principles of electrophysiology

The underlying theme of this monograph is that the formal description of *all* transport phenomena centers about the determinations of flows and forces that are then related by means of generalized conductances or resistances. Up to this point we have focused on the relations between fluxes (determined using chemical or isotopic techniques), electrochemical potential differences, and permeabilities, that is, the kinetic approach. In this final chapter we will briefly consider the electrophysiologic approach *toward the same end*, for two reasons. First, whereas many students of the field are comfortable with the concepts of fluxes, concentration differences, electrical potential differences, and permeabilities, they often react with horror when faced with equivalent electrical circuits that involve currents, resistors, and electromotive forces (batteries); our aim will be to show that these are in fact two ways of expressing the very same concepts. Second, whereas the kinetic and electrophysiologic approaches share the same theoretical base and differ only in terminology, the latter enjoys a very advanced technology and, as a result, a number of practical advantages. For example, currents as low as 10^{-12} A can be measured accurately with almost instantaneous time resolution; this is equivalent to the net flux of approximately 1×10^{-17} equivalents of an ion and there is no chemical or isotopic technique that can come close to matching this level of resolution. There are a number of other technical advantages offered by the electrophysiologic approach that need not be belabored; suffice it to say that the brilliant studies of Hodgkin, Huxley, Katz, Cole, and others that led to the unraveling of many of the mysteries of the axon (Cole, 1968) could not have been accomplished using other approaches.[1]

7.1. Electrical circuit analogs of flows and forces

Let us begin by assuming that the flow of an ion i is strictly diffusional so that, in general, it can be described by the Nernst-Planck equation

$$J_i = -u_i\bar{c}_i z_i \mathcal{F}[(RT/z_i\mathcal{F})(d \ln \bar{c}_i/dx) + (d\bar{\psi}/dx)] \qquad (7.1)$$

As discussed earlier, the electrical current resulting from this flow is given by $I_i = z_i \mathcal{F} J_i$, so that equation (7.1) can be rearranged to give

$$\frac{I_i}{u_i z_i^2 \mathcal{F}^2 \bar{c}_i} = -\left(\frac{RT}{z_i \mathcal{F}}\right)\left(\frac{d \ln \bar{c}_i}{dx}\right) - \left(\frac{d\bar{\psi}}{dx}\right) \tag{7.2}$$

Assuming that I_i is constant (steady state), equation (7.2) can be integrated over the thickness of the membrane yielding

$$I_i \int_o^i \left(\frac{dx}{u_i z_i^2 \mathcal{F}^2 \bar{c}_i}\right) = \left(\frac{RT}{z_i \mathcal{F}}\right) \ln\left(\frac{c_i^o}{c_i^i}\right) - \Delta\psi \tag{7.3}$$

where, as before, $\Delta\psi = \psi^i - \psi^o$. It should be noted that the right-hand side of equation (7.3) refers to properties of the external solutions; as discussed previously, this is valid providing $\Delta\psi$ includes boundary (interfacial) electrical potential differences.

Now, following Finkelstein and Mauro (1963), we define the *integral resistance* of the membrane to i as

$$R_i = \int_o^i (dx/u_i z_i^2 \mathcal{F}^2 \bar{c}_i) \tag{7.4}$$

and the *electromotive force* of i as the Nernst potential,

$$E_i = (RT/z_i \mathcal{F}) \ln(c_i^o/c_i^i) = -(\Delta\mu_i/z_i \mathcal{F}) \tag{7.5}$$

Substituting equations (7.4) and (7.5) into equation (7.3), we obtain

$$I_i R_i = E_i - \Delta\psi \tag{7.6}$$

Equation (7.6) can be readily recognized as a statement of Ohm's law and, more generally, is an example of the linear phenomenological equation that relates flows and forces (1.33). In essence, this equation states that the Nernst–Planck equation for the flow of an ion driven by its electrochemical potential difference can be represented (symbolized) by a current I_i flowing across a resistor R_i that is connected in series with a battery E_i as illustrated in Figure 7.1(a); the *measured* electrical potential difference across the barrier is obviously given by

$$\psi^i - \psi^o = \Delta\psi = E_i - I_i R_i$$

Clearly, equation (7.4) cannot be integrated without some assumptions regarding u_i and the concentration profile of i across the membrane (i.e., the relation between $\bar{c}_i$ and x). However, insight into the meaning of R_i can be gained by considering the case where u_i is constant and $c_i^o \cong c_i^i$, so that $\bar{c}_i$ in equation (7.5) can be replaced with an average concentration $\dot{c}_i$ that is equal

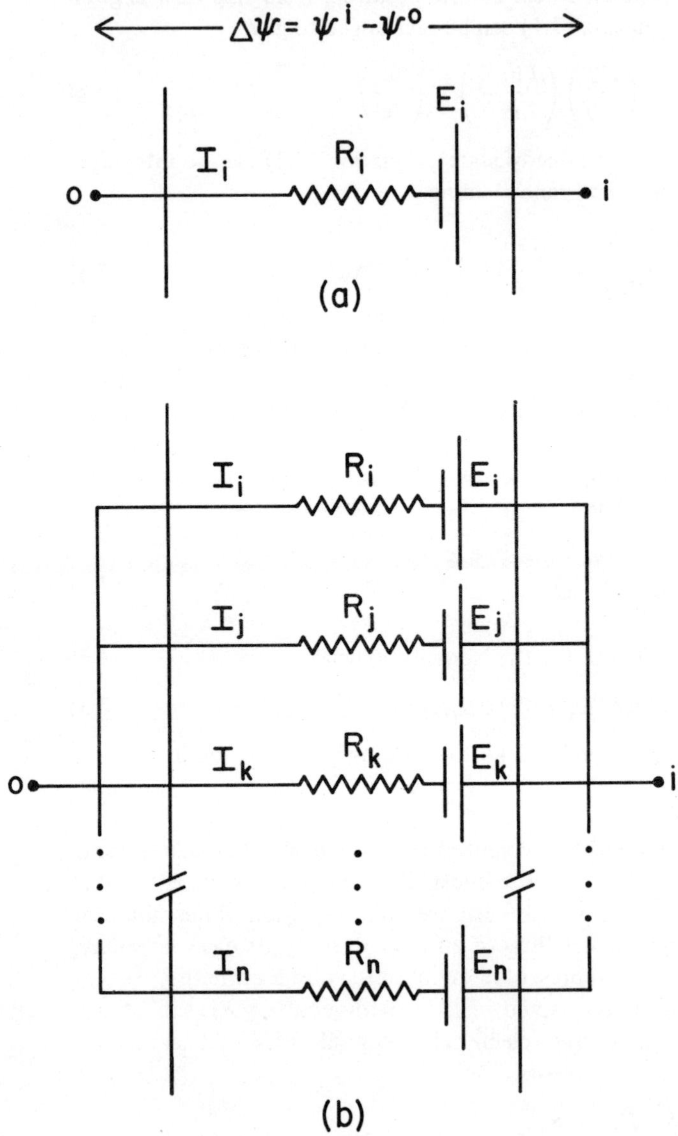

Figure 7.1. (a) Equivalent electrical circuit representation of the Nernst–Planck equation for ionic diffusion. (b) Equivalent electrical circuit for the diffusion of n ions across a membrane. (*Note:* As discussed in Section 7.2, the electromotive force E can be generalized to include flows that are not accounted for by simple diffusion.)

to $[\beta_i(c_i^o + c_i^i)/2]$. Then,

$$R_i = [\Delta x/u_i z_i^2 \mathcal{F}^2 \dot{c}_i] \tag{7.7}$$

Since, $P_i = \beta_i RTu_i/\Delta x$, the relation between R_i and P_i is

$$R_i = [RT/P_i z_i^2 \mathcal{F}^2 \dot{c}_i] \tag{7.8}$$

The conductance of the membrane to i is, simply, $G_i = (1/R_i)$ so that

$$G_i = [P_i z_i^2 \mathcal{F}^2 \dot{c}_i/RT] \tag{7.9}$$

Two important points should be noted:

First, the electrical resistance of the membrane to i is directly related to its thickness and inversely related to the local concentrations of i integrated over the membrane thickness, or the *content* of i per unit area.[2]

Second, although G_i is directly related to P_i, these terms are *not* synonymous inasmuch as G_i is also dependent upon $\dot{c}_i$. This point, which will be illustrated below, must be emphasized since the terms *conductance* and *permeability* are often used interchangeably in the literature.

Now, if we assume that the electric field within the membrane is constant, equation (7.1) can be integrated directly, yielding

$$I_i = -\left(\frac{P_i z_i^2 \mathcal{F}^2 \Delta\psi}{RT}\right)\left[\frac{c_i^o - c_i^i \exp(z_i \mathcal{F}\Delta\psi/RT)}{1 - \exp(z_i \mathcal{F}\Delta\psi/RT)}\right] \tag{7.10}$$

Further, from equation (7.6) we see that when E_i is constant (i.e., when c_i^o and c_i^i are constant), $R_i = -[\partial(\Delta\psi)/\partial I_i]$, so that

$$R_i = -\left[\frac{(RT)^2}{P_i z_i^3 \mathcal{F}^3 \Delta\psi}\right]\left[\frac{(1-\xi)\ln(c_i^o/c_i^i\xi)}{c_i^o - c_i^i\xi}\right] \tag{7.11}$$

where $\xi = \exp(z_i \mathcal{F}\Delta\psi/RT)$.

Clearly, multiplying equation (7.10) by equation (7.11) yields equation (7.6).

When $c_i^o \cong c_i^i$ then $I_i \cong -[(P_i z_i^2 \mathcal{F}^2 \Delta\psi\dot{c}_i)/RT]$ and R_i approaches the expression given by equation (7.8).

Further, when $\Delta\psi = 0$, then $I_i = P_i z_i \mathcal{F}(c_i^o - c_i^i)$ and

$$R_i = \left(\frac{RT}{P_i z_i^2 \mathcal{F}^2}\right)\left[\frac{\ln(c_i^o/c_i^i)}{(c_i^o - c_i^i)}\right] \tag{7.12}$$

As discussed previously (Section 4.2), the term $[\Delta c_i/\ln(c_i^o/c_i^i)]$ is the *logarithmic mean* concentration of i, which approaches the *arithmetic mean* $\dot{c}_i$ when $\Delta c_i \to 0$, so that equation (7.12) approaches equation (7.8).

We now return to an issue that was introduced earlier (Section 2.3) and deferred for further discussion, namely, "Goldman-type" rectification. Referring to Figure 2.3, we see that according to the constant field equation a plot of J_i versus $\Delta\psi$ is linear only when $c_i^o = c_i^i$; when $c_i^o \neq c_i^i$, the plot is curvilinear in spite of the fact that P_i, c_i^o and c_i^i are constant. Now, the ordinate labeled J_i can be relabeled I_i $(=z_i\mathcal{F}J_i)$ so that the curves actually are current-voltage or "*I-V*" plots. The slopes of these curves, at any value of $\Delta\psi$ are given by $[\partial I_i/\partial(\Delta\psi)]$ (the partial differential is employed to indicate that P_i, c_i^o, and c_i^i are maintained constant) and are equal to G_i (or $1/R_i$). Clearly, when $c_i^o \neq c_i^i$, *the slope conductance* G_i is a function of $\Delta\psi$ and is lower when I_i is directed from the compartment with the lower concentration to that with the higher concentration (in the example shown, from compartment *i* to compartment *o* so that I_i is negative) than when I_i is in the opposite direction. When $c_i^o = c_i^i$, G_i is independent of $\Delta\psi$ and the relation between I_i and $\Delta\psi$ is linear.

Figure 7.2 is a plot of R_i versus $\Delta\psi$ for the conditions illustrated in Figure 2.3, calculated using equation (7.11); these curves clearly confirm the impressions gained from inspection of Figure 2.3. Thus, when $c_i^o = 100$ mM and $c_i^i = 10$ mM, R_i decreases as $\Delta\psi$ is varied from +100 to -100 mV; when $c_i^i = 50$ mM, R_i is always less than when $c_i^i = 10$ mM for any finite value of $\Delta\psi$ but still decreases when $\Delta\psi$ is varied from +100 to -100 mV; when $c_i^o = c_i^i = 100$ mM, R_i is still lower and is now independent of $\Delta\psi$ [see equation (7.8)].

Figure 7.3 provides the explanation for these phenomena. This figure illustrates the concentration profiles of *i* within the membrane under several conditions calculated according to Arndt and Roper (1972).

When $c_i^o = 100$ mM, $c_i^i = 10$ mM and $\Delta\psi = +100$ mV, the concentration profile is concave upward; since I_i is directed from compartment *i* to compartment *o*, this profile can be viewed as the result of a convective (fluid) flow of current, interacting with *i*. Conversely, when $\Delta\psi = -100$ mV, I_i is directed from compartment *o* to compartment *i* and the concentration profile is convex upward. When $\Delta\psi = 0$, the concentration profile is linear. It is obvious that the *amount* of *i* within the membrane (per unit area) is least when $\Delta\psi = 100$ mV, most when $\Delta\psi = -100$ mV and intermediate when $\Delta\psi = 0$. From equation (7.4) we see that R_i is inversely related to the amount of (mobile) *i* within the membrane. It follows that R_i will be greater when $\Delta\psi = 100$ mV than when $\Delta\psi = -100$ mV and that an intermediate value of R_i will be observed when $\Delta\psi = 0$.

When $c_i^o = c_i^i = 100$ mM, the concentration of *i* within the membrane is independent of position and $\Delta\psi$; consequently, R_i is independent of $\Delta\psi$. Further, it is obvious that the amount of *i* within the membrane under this condition (for any finite value of $\Delta\psi$) is always greater than that when $c_i^i = 10$ mM so that R_i will always be lower (see Figure 7.2).

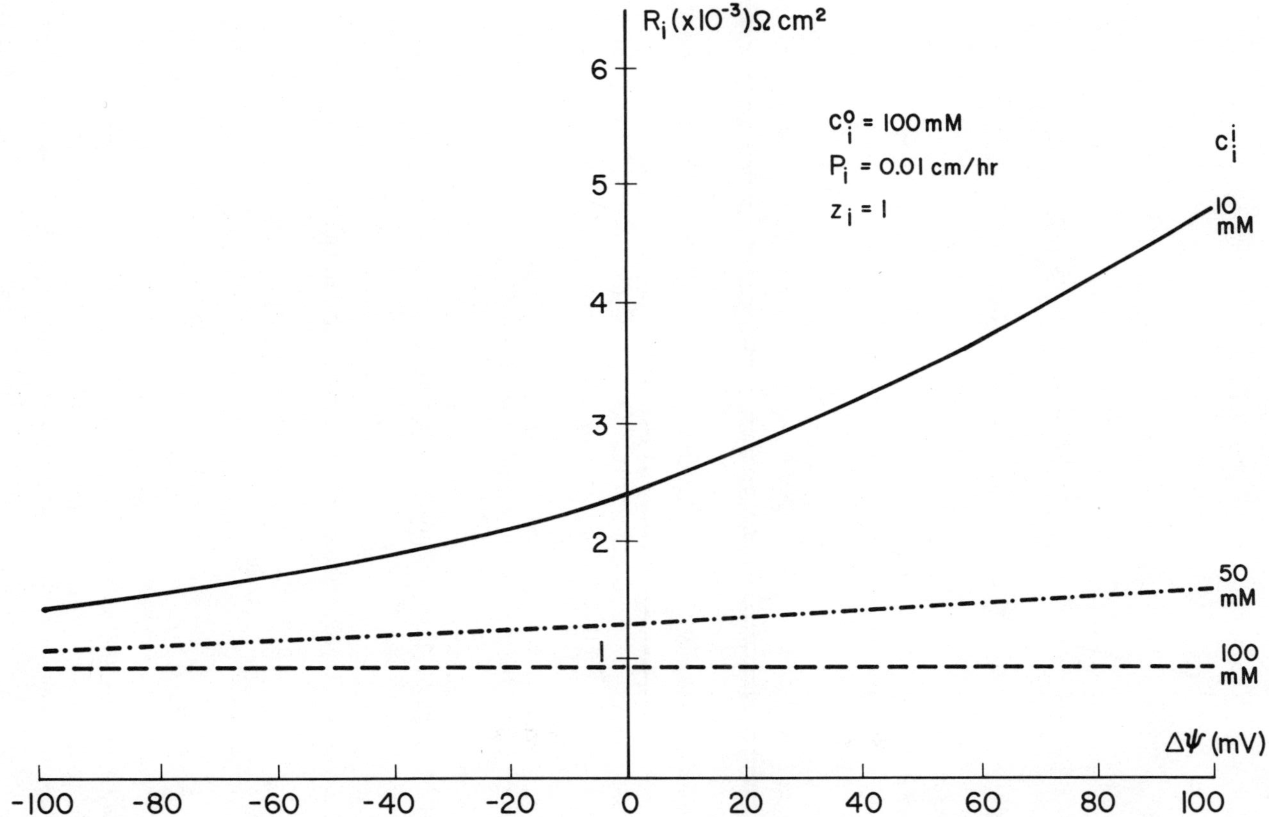

Figure 7.2. R_i as a function of $\Delta\psi$ for a monovalent cation when $c_i{}^o$ and P_i are maintained constant and $c_i{}^i$ is varied.

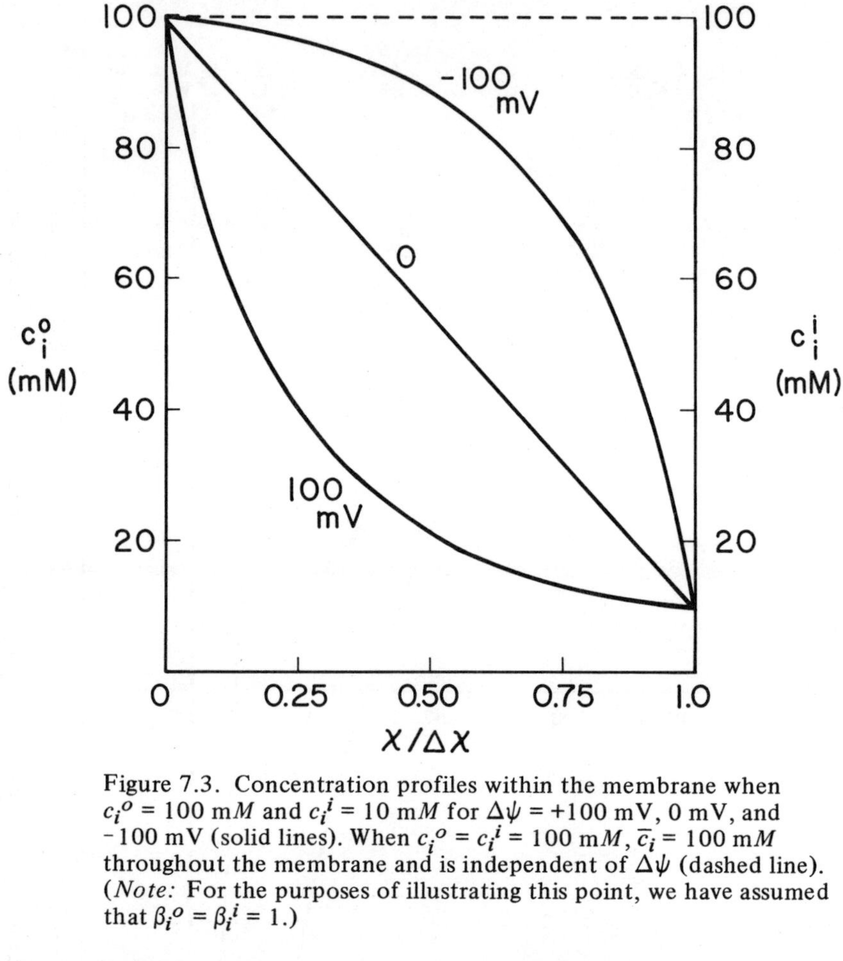

Figure 7.3. Concentration profiles within the membrane when $c_i^o = 100$ mM and $c_i^i = 10$ mM for $\Delta\psi = +100$ mV, 0 mV, and -100 mV (solid lines). When $c_i^o = c_i^i = 100$ mM, $\bar{c}_i = 100$ mM throughout the membrane and is independent of $\Delta\psi$ (dashed line). (*Note:* For the purposes of illustrating this point, we have assumed that $\beta_i^o = \beta_i^i = 1$.)

7.2. The "overall" resistance and electromotive force

If there are parallel flows of n ions across a membrane, equation (7.6) can be written as follows:

$$I = \sum_i^n I_i = \sum_i^n G_i E_i - \Delta\psi \sum_i^n G_i \tag{7.13}$$

Now, in this parallel system, the total conductance of the membrane G_m is equal to $\Sigma_i^n G_i$, so that equation (7.13) reduces to

$$IR_m = \sum_i^n (G_i E_i / G_m) - \Delta\psi \tag{7.14}$$

where R_m is the *total electrical resistance* of the membrane and is given by

$$R_m = \left[\sum_i^n (1/R_i)\right]^{-1} = (1/G_m)$$

G_i is referred to as the *partial ionic conductance* of i, inasmuch as it represents the contribution of I_i to the total membrane conductance.

Clearly, from equations (7.6) and (7.13) it follows that $G_i = [\partial I_i/\partial(-\Delta\psi)]$ and $G_m = [\partial I/\partial(-\Delta\psi)]$ so that $(G_i/G_m) = (\partial I_i/\partial I)$. In other words, (G_i/G_m) is the ratio between the change in I_i to the change in total current I in response to an imposed $\Delta\psi$ when all E_i (i.e., c_i^o and c_i^i) are maintained constant; as discussed earlier, this is the *transference number* t_i of the *total* membrane to the species i under the given set of conditions.[3] Thus, equation (7.14) can be rewritten as follows:

$$IR_m = E - \Delta\psi \tag{7.15}$$

where E is the *total electromotive force* across the membrane given by

$$E = \sum_i^n t_i E_i \tag{7.16}$$

If the ionic flows are strictly diffusional, E_i is given by equation (7.5) which, when substituted into equations (7.15) and (7.16), yields

$$IR_m = \frac{RT}{\mathcal{F}} \sum_i^n \left(\frac{t_i}{z_i}\right) \ln\left(\frac{c_i^o}{c_i^i}\right) - \Delta\psi \tag{7.17}$$

Thus, when $I = 0$

$$\Delta\psi = \frac{RT}{\mathcal{F}} \sum_i^n \left(\frac{t_i}{z_i}\right) \ln\left(\frac{c_i^o}{c_i^i}\right) = -\frac{1}{\mathcal{F}} \sum_i^n \left(\frac{t_i}{z_i}\right) \Delta\mu_i \tag{7.18}$$

Equation (7.18) is identical to equation (3.18) which describes the multi-ionic diffusion potential under "zero current conditions." In other words, E is the electrical potential difference across the membrane under zero current conditions *or* the electrical potential difference at which the membrane must be clamped in order to abolish current flow.

Clearly, equation (7.17) can be represented by the equivalent electrical circuit illustrated in Figure 7.1(b), where, in general, $R_i, R_j, R_k, \ldots, R_n$ and $E_i, E_j, E_k, \ldots, E_n$ are given by equations (7.4) and (7.5); and $I_i, I_j, I_k, \ldots, I_n$ are given by equation (7.6). In the particular case where the electric field within the membrane is constant, $R_i, R_j, R_k, \ldots, R_n$ are given by equation (7.11), $I_i, I_j, I_k, \ldots, I_n$ are given by equation (7.10), and $E_i, E_j, E_k, \ldots, E_n$ are given by equation (7.5).

Finally, the expression for the overall electromotive force can be generalized to include the possible effects of nonconjugate driving forces on I_i. In short, any process that can generate a flow (current) of i will contribute to E_i and E. These include the coupling of J_i to solvent flow (solvent drag); the coupling of J_i to the flow of another solute, j (solute-solute cotransport, solute-drag); the coupling of J_i to a metabolic reaction (active rheogenic transport); and so on. Thus, in general,

$$
\begin{aligned}
E = & -\left(\frac{1}{\mathfrak{F}}\right)\sum_i^n \left(\frac{t_i}{z_i}\right) \Delta\mu_i && \text{(Diffusion potential)} \\
& + \frac{1}{G_m}\sum_i^n z_i\mathfrak{F}(1 - \sigma_i)\, \bar{c}_i J_v && \text{(Streaming potential)} \\
& + \frac{1}{G_m}\sum_i^n z_i\mathfrak{F}\nu J_j && \text{(Cotransport potential)} \\
& + \frac{1}{G_m}\sum_i^n z_i\mathfrak{F}\,{}_aJ_i + \cdots && \text{(Active transport potential)}
\end{aligned}
\tag{7.19}
$$

where ν is the coupling coefficient between J_i and J_j and ${}_aJ_i$ is the flow of i resulting from active transport.

Thus, the equations of flow that have been discussed in previous sections can be translated into measurable electrical parameters such as currents, resistances, batteries (electromotive forces), and electrical potential differences and equations (7.15) and (7.19) are simply the Thévenin equivalents of a system characterized by the flow of one or more ions; Figure 7.1(b) is the electrical circuit representation of such a system.

7.3. The "short-circuit technique"

Although the emphasis in this monograph has been on principles rather than techniques, the short-circuit technique has such instructive as well as practical value that it would be remiss if it were not discussed in some detail. This technique was introduced by Ussing and Zerahn (1951) for the study of Na transport by isolated frog skin but, whereas it has been employed almost exclusively for the investigation of ion transport by epithelial tissues, in principle, it is applicable to any membrane.

Let us first consider the theoretical basis of this technique and then illustrate its application.

If a membrane is bathed on both surfaces by a solution of a salt that dissociates into a cation i with valence z_i and an anion j with a valence z_j, the system can be represented by the equivalent circuit illustrated in Figure 7.4.

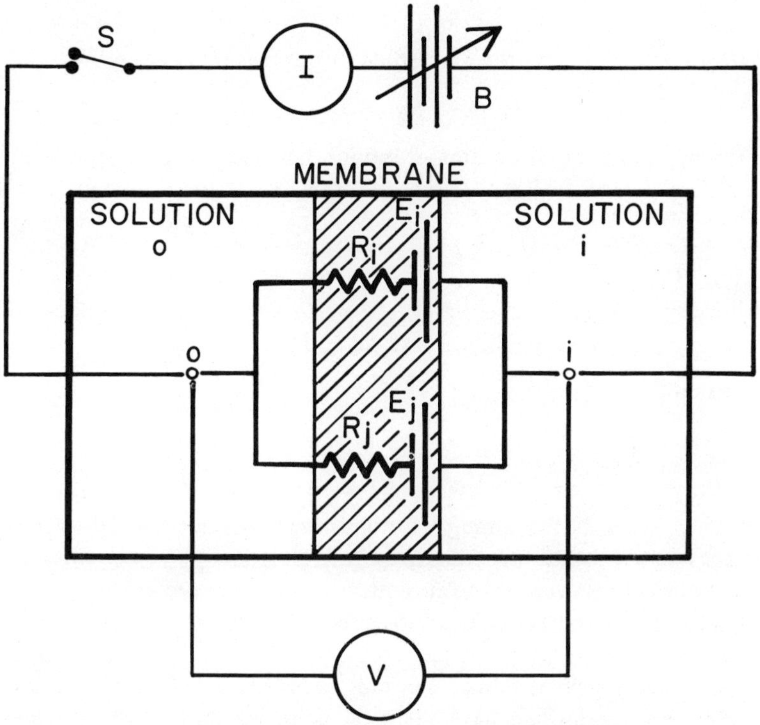

Figure 7.4. Equivalent electrical circuit for the transport of a cation i and anion j across a membrane under open-circuit and short-circuit conditions.

As shown, provisions are made to measure the electrical potential difference across the membrane using the voltmeter V (which is assumed to draw no current) and to pass an external current I across the barrier driven by a variable battery B.

The general solution of this circuit is

$$IR_m = E - \Delta\psi$$

where $R_m = R_iR_j/(R_i + R_j)$ and $E = R_m[(E_i/R_i) + (E_j/R_j)]$ or alternatively [equation (7.16)]

$$E = (G_i/G_m)E_i + (G_j/G_m)E_j$$

When $I = 0$ (i.e., the switch S is open)

$$\Delta\psi = E = R_m[(E_i/R_i) + (E_j/R_j)] \quad (7.20)$$

Thus, assuming that R_i and R_j are finite, there will be an electrical potential difference across the membrane if E_i and/or E_j are nonzero; this is referred to as the *open circuit* potential. This electrical potential difference can be abolished by passing an opposing current across the membrane such that $IR_m = -\Delta\psi$. Clearly, the magnitude of this current is given by

$$(I)_{\Delta\psi=0} = (E_i/R_i) + (E_j/R_j) \tag{7.21}$$

Now, when both bathing solutions have identical compositions so that $c_i^o = c_i^i = c_i$ and $c_j^o = c_j^i = c_j$, equations (7.19) and (7.20) yield (when $I = 0$)

$$\Delta\psi = R_m[(E_i/R_i) + (E_j/R_j)] = R_m z_i \mathfrak{F}(J_i)_{\Delta c_i=0} + R_m z_j \mathfrak{F}(J_j)_{\Delta c_j=0}$$

or

$$\Delta\psi = R_m\,[(I_i)_{\Delta c_i=0} + (I_j)_{\Delta c_j=0}] \tag{7.22}$$

so that

$$(I)_{\Delta c,\,\Delta\psi=0} = \frac{\Delta\psi}{R_m} = [(I_i)_{\Delta c_i=0} + (I_j)_{\Delta c_j=0}] \tag{7.23}$$

Thus, when both bathing solutions have identical compositions, the current necessary to abolish the transmembrane electrical potential difference is equal to the *algebraic sum* of the currents of *i* and *j that are driven by nonconjugate forces;* this current is referred to as the *short-circuit current*, which we denote I_{sc}.

In other words, if *i* and *j* can only cross the membrane by diffusion, then, when both solutions have identical compositions, there will be no flows, $E = 0$ (7.19), $\Delta\psi = 0$ (7.18), and obviously $I_{sc} = 0$ (7.21). However, if *i* and/or *j* are driven by nonconjugate forces, then, when $\Delta c_i = \Delta c_j = 0$, in general, $E \neq 0$, $\Delta\psi \neq 0$, and the I_{sc} is given by the algebraic sum of the currents of *i* and *j*.

In order to illustrate these points further, let us consider the system shown in Figure 7.4, assuming that the anion *j* can only cross the membrane by simple diffusion and that the cation *i* is actively transported from compartment *o* to compartment *i*. When $c_i^o = c_i^i$ and $c_j^o = c_j^i$, then $E_j = 0$, and from equation (7.22) under open-circuit conditions (i.e., when $I = 0$),

$$\Delta\psi = E = R_m E_i/R_i \tag{7.24}$$

Since $E = R_m \mathfrak{F} z_{i\,a}J_i$ [equation (7.19)], it follows that $\Delta\psi > 0$; that is, the inner solution will be electrically positive with respect to the outer solution.

Now, what are the individual currents through the two parallel pathways under open-circuit conditions? These may be described as follows

$$I_i R_i = E_i - \Delta\psi \tag{7.25}$$

and

$$I_j R_j = -\Delta\psi \tag{7.26}$$

Thus, the flow of j from the outer to the inner solution is driven solely by the electrical potential difference established by the active transport of i. Clearly, in order that electroneutrality be maintained under open-circuit conditions, the restriction on the currents is that $I_i = -I_j$. Using this restriction and equations (7.25) and (7.26), we obtain equation (7.24). Thus, the physical system is *consistent with* the equivalent electrical circuit model; the reason for the emphasis on "consistent with" will be discussed in Section 7.4.

Now, the amount of an (opposing) current that must be passed across the membrane from the external source B in order to clamp $\Delta\psi$ to zero is simply

$$I_{sc} = (E/R_m) = (E_i/R_i) = z_i \mathcal{F}\, {}_aJ_i \qquad (7.27)$$

Thus, the short-circuit current is equal to the rate at which i is actively transported from compartment o to compartment i, when ${}_aJ_i$ (in moles per centimeter squared, second) is converted into amperes per centimeter squared by the factor $z_i\mathcal{F}$.

Finally, it should be pointed out that when $\Delta\psi = 0$, there is no current flow through R_j so that the entire I_{sc} passes through the pathway responsible for active transport of i but in the opposite direction. Thus, the I_{sc} may be viewed as the external current that must be passed through the membrane in order to "buck out" the current generated within the membrane by the active transport mechanism and thereby render $\Delta\psi = 0$.[4]

Now let us descend from theory to practice and examine the application of the short-circuit technique to a "real" membrane where Na is actively transported from the outer to the inner solution and the movements of Cl are entirely diffusional; this system is illustrated in Figure 7.5. As already discussed, when both surfaces of the membrane are bathed with identical NaCl solutions, under open-circuit conditions the inner solution will be electrically positive with respect to the outer solution and

$$\Delta\psi = R_m \mathcal{F}\, {}_aJ_{\mathrm{Na}}$$

Now, since electroneutrality must be maintained, there must be a flow of Cl from the outer solution to the inner solution equal to ${}_aJ_{\mathrm{Na}}$. Since $[\mathrm{Cl}]^o = [\mathrm{Cl}]^i$, this flow must be driven entirely by $\Delta\psi$; thus, the electrical potential difference secondary to the active transport of Na is the driving force for the accompanying flow of Cl. The magnitude of $\Delta\psi$ will be directly related to the rate of active Na transport and the resistance of the tissue to Cl diffusion. [The resemblance of this line of reasoning to that underlying the diffusion potential between two solutions of the same salt (Section 3.1) should be noted. In that case, $\Delta\psi$ was determined by the difference between anion and cation mobilities and the magnitude of the concentration ratio. In this case, there is no concentration difference but $\Delta\psi$ is determined by properties of the system related to the maintenance of electroneutrality.]

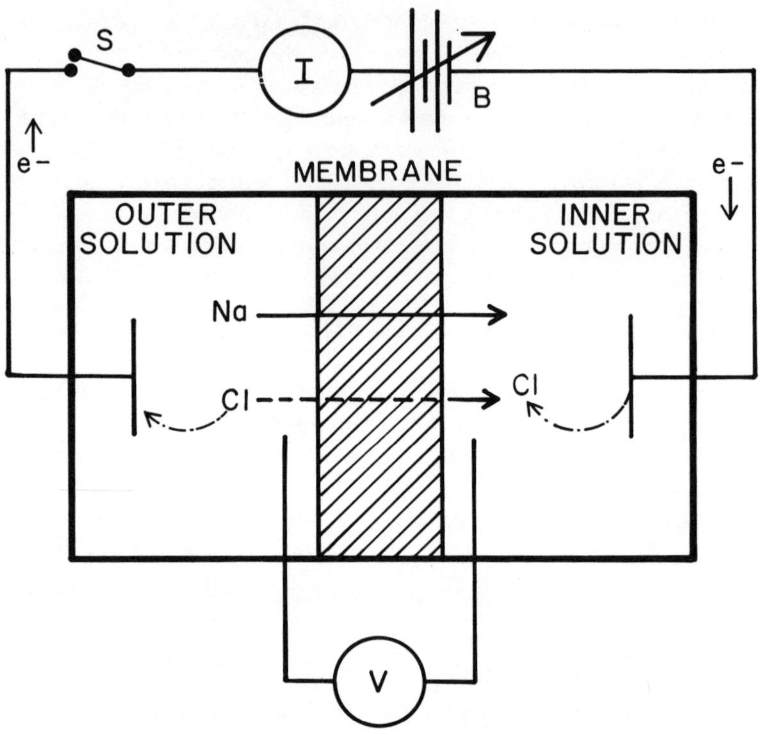

Figure 7.5. Illustration of the application of the short-circuit technique to the study of a membrane that is capable of actively transporting Na from the outer to the inner solution; Cl transport is assumed to be entirely diffusional.

When sufficient current is passed across the membrane to abolish the open circuit $\Delta\psi$, clearly, $\Delta\tilde{\mu}_{\text{Na}} = \Delta\tilde{\mu}_{\text{Cl}} = 0$. This is referred to as the *short-circuit* condition, inasmuch as the electrical potential difference across the membrane has been "shorted"; it is as if the membrane were bypassed with an infinitely high conductor (infinitely low resistor). Because Na is actively transported, a net flow of Na from the outer to the inner solution will persist in spite of the fact that $\Delta\tilde{\mu}_{\text{Na}} = 0$. On the other hand, because the flow of Cl is entirely diffusional, J_{Cl} will cease when $\Delta\tilde{\mu}_{\text{Cl}} = 0$.

How, then, is electroneutrality maintained under short-circuit conditions?

Let us consider how an external current is passed across a membrane, surrounded by electrolyte solutions, from an electromotive force (the battery) via electrodes immersed in the two solutions. In solid-state or metallic conductors, current is carried by electrons, but in solutions it is carried by ions (electrolytic conduction). It follows that at the interfaces between the current-passing electrodes and the solutions in Figure 7.5, processes take place that convert ions into electrons and vice versa; these are referred to as *electrode reactions.* Now, let us assume that the current-passing electrodes consist of silver wires or plates that have been electroplated with Cl, that is, silver-silver chloride (Ag/AgCl) electrodes. At the cathode the electrode reaction is

$$AgCl + e^- \longrightarrow Ag + Cl^-$$

so that an electron driven down the wire from the negative pole of the battery generates a Cl ion that enters the solution. At the anode, the opposite reaction takes place, that is,

$$Cl^- + Ag \longrightarrow AgCl + e^-$$

so that a Cl ion is "plated out" of solution; the electron released is drawn to the positive pole of the battery and the circuit is completed!

Thus, when sufficient current is passed across the tissue to abolish $\Delta\psi$ and render $\Delta\tilde{\mu}_{Cl} = 0$, Cl transport *through the membrane* ceases but there is a net displacement of Cl from the outer to the inner solution resulting from the electrode reactions. In essence, for every Na actively transported across the membrane from compartment o to compartment i, one Cl in the outer solution (which contains the anode) is plated out and the electron that is released flows through the ammeter to the positive "sink" of the battery. At the same time an electron derived from the negative "source" of the battery interacts with AgCl at the cathode resulting in the release of Cl ion into the inner solution. Thus, the rate of flow of electrons through the external circuit is precisely equal to the flow of Na from the outer to the inner solution through the membrane, so that the I_{sc} is equal to the active Na current and electroneutrality is maintained.

Obviously, the same reasoning can be applied to a membrane capable of actively transporting several ions having different valences in different directions. It is a simple matter to show that the I_{sc} (i.e., the rate of electron flow in the external circuit) is always the algebraic sum of the ionic currents through the membrane.

7.4. A caveat regarding electrical circuit models

In their important paper dealing with the use of equivalent electrical circuits to model ionic movements across membranes, Finkelstein and Mauro (1963) point out a significant shortcoming of this approach that must be kept in mind. Briefly, the problem arises from the fact that Kirchhoff's laws, which are used to solve electrical circuits, deal exclusively with the flow of electrons but do not recognize different ionic species, that is, these laws do not discriminate between an electron carried by Cl and one carried by HCO_3 nor do they discriminate between the current carried by the flow of a monovalent anion in one direction and that carried by the equal flow of a monovalent cation in the opposite direction; these are *electrically* equivalent. *But, artificial and biological membranes do make these distinctions!*

For example, in Figure 7.1(b), there is no formal way of preventing the current I_i from flowing across the resistor R_j; according to Kirchhoff's laws, currents are currents and resistors are resistors and the subscripts $i, j, k, \ldots, n$ that designate different ionic species are unrecognized. With reference to Figure 7.4, we employed equations (7.25) and (7.26) to *force* I_i through R_i and I_j through R_j and then imposed the physical restriction that $I_i = -I_j$; these impositions are not natural consequences of a formal solution of the equivalent electrical circuit model.

Some of the possible misleading consequences of applying equivalent electrical circuit models to multi-ionic systems have been discussed by Finkelstein and Mauro (1963) for single plasma membranes and by Schultz and others (1977) and Schultz (1979) for epithelial tissues. The purposes of these caveats are not to disparage or discourage the use of this valuable approach to modeling electrophysiologic properties of membranes. Instead, it must be emphasized that they must be employed with circumspection and that, often, restrictions must be placed on the model in order to transform a collection of wires, batteries, and resistors into a realistic portrayal of a biological system.

Notes

Chapter 1

1. There are many excellent texts on classical thermodynamics. My favorites are Lewis and Randall (1961), Glasstone (1956), Denbigh (1966), and Guggenheim (1950), the latter being a very rigorous treatment. For an unusual and delightful approach see Angrist and Hepler (1967).
2. We will emphasize *closed systems* (i.e., those that do not exchange matter with their surroundings) at constant temperature.
3. The partial differential $(\partial x/\partial y)_{u,v,w}$ is the derivative of x with respect to y when other variables are maintained constant. Thus, the integral of $[(\partial x/\partial y)_{u,v,w}]\, dy$ from an initial state o to a final state i is the *contribution* of the change in *y alone* to the total change in x.

   In some instances for the sake of clarity we will omit the subscripts $u, v, w, \ldots$ and simply write $(\partial x/\partial y)$.
4. We now refer to a "closed system" that can exchange heat but not matter with its surroundings. As will be discussed briefly, matter also "contains" entropy so that in an "open system" changes in S, in general, result from exchanges of heat and matter with the surroundings.
5. It is a common misconception that, in general, a difference in temperature is the driving force for heat flow. This is true only for a "closed system" which cannot exchange matter with its surroundings. Since matter has a heat capacity, a difference in temperature is a driving force for the flow of matter [see equation (1.29)] and the exchange of entropy between an "open system" and its surroundings involves both the flow of heat and the flow of matter.

   Further, it should be noted that P in equation (1.5) now refers to the internal pressure of the system rather than the external pressure of equation (1.2), since if the process is reversible, the internal and external pressures can only differ infinitesimally.
6. Josiah Willard Gibbs (1839–1903), a chemist, as well as a theoretical physicist, was particularly interested in reactions that take place at constant temperature and pressure and was not concerned with "work" resulting from a change in volume of a solution undergoing a chemical reaction; hence the implication that this (PdV) work is "useless."
7. Clausius coined the term *entropy* which stems from the Greek, *entrope*–the act of turning toward or turning into. He clearly recognized that this function reflects the "directionality" of spontaneous

processes and is a measure of the degradation of a system's ability to perform work.

8. Boltzmann's statistical mechanical treatment of the second law made the increase in entropy which accompanies an irreversible process a *highly probable* occurrence rather than the absolute certainty envisioned by Clausius. Planck initially rejected this probabilistic approach and undertook his studies on "black-body" radiation, in part to establish the *absoluteness* of the concept of irreversibility. In the end, he was forced to accept the statistical basis of the second law and his studies, which were designed for an entirely different purpose, led to the quantum theory (M. J. Klein, *Physics Today*, 19:23–32, 1966).
9. It must be emphasized that all *potentials* (intensive properties) must be referred to an arbitrary "reference point," or "ground state." Thus, when we refer to a pressure, it is generally with respect to that at sea level which is *assigned* the value 0 mm Hg. The electrical potential at a point is with respect to "ground," which is assigned the value 0 volts. And, as is well known, temperature on the centigrade scale is referred to the freezing point of water (at atmospheric pressure), or on the Kelvin scale to the hypothetical point where the volume of an ideal gas vanishes (−273.16 °C). A difference between two pressures is meaningful only when each pressure has the same reference point; the same is true for a difference in electrical potential and a difference in temperature. Similarly, there is no such thing as an absolute electrochemical potential. All values of $\tilde{\mu}_i$ are with reference to a "standard (ground) state," where $T = 0, P = 0, \psi = 0$, and $c_i = 1$ in a given solvent; clearly, under these conditions $\tilde{\mu}_i = \mu_i^0$. A difference in electrochemical potential difference is meaningful only when reference is made to the same standard (ground) state.
10. For general references see DeGroot (1952), DeGroot and Mazur (1962), Katchalsky and Curran (1965) and Prigogine (1961); each is a classic in its own right.

Chapter 2

1. Thus, J_i is defined as positive when directed from compartment o to compartment i; this convention will be strictly adhered to throughout. It should be noted that for charged species the current flow attributable to J_i is $z_i \mathcal{F} J_i$, so that for anions the current I_i is negative when J_i is positive.
2. For a discussion of the effects of unstirred layers see Lakshminarayanaiah (1969) or Helfferich (1962).
3. We will only be concerned with constant or "steady-state" velocities. Under these conditions the total force acting on a particle is zero, so that the driving force is "balanced" by retarding, frictional forces; the mobility is a measure of these frictional forces. For example, for diffusion of a spherical particle with radius r_i in a fluid whose viscosity is η, $v_i = (1/6\pi\eta r_i) f_i$. Thus, u_i is inversely proportional to the viscosity of the surrounding fluid and the radius of the particle (the Stokes–Einstein relation). (See L. G. Longsworth in *Electrochemistry in biology and medicine* (T. Shedlovsky, ed.), New York: John Wiley, 1955, pp. 225–47).

4. For a solute with a partial molar volume of 20 cm^3/mole in a solution at atmospheric pressure, $\overline{v}_i P = 20(cm^3/mole) \times 10^6$ (dynes/cm^2) = 2 J/mole. In contrast, RT at 25°C ≅ 2500 J/mole, and for a monovalent cation $z_i \mathcal{F} \psi = 96{,}500\psi$ J/mole when ψ is expressed in volts.
5. It is generally assumed that i is distributed at equilibrium across each interface. In essence this implies that diffusion through the membrane is slow compared to partitioning across the interfaces. Under these conditions we can write *for each interface* $\tilde{\mu}_i = \overline{\tilde{\mu}}_i$. From equation (1.28) it follows that when $z_i = 0$, $\beta_i = \overline{c}_i/c_i = \exp[(\mu_i{}^0 - \overline{\mu}_i{}^0)/RT]$. Thus, β_i is a function of the difference in standard state free energy, which (at constant T and P) is a function of the different "solvent properties" of the membrane and the adjacent solution(s). If the two have the same solvent properties for i, $\beta_i = 1$. On the other hand, if i is a lipophilic molecule, the outer solutions are aqueous, and the membrane is lipid, i will be more "stable" in the membrane phase; accordingly, $\overline{\mu}_i{}^0 < \mu_i{}^0$ and $\beta_i > 1$.
6. Equation (2.21) can be derived from more general considerations when $z_i \mathcal{F} \Delta\psi < RT$ (Parlin and Eyring, 1954; Kimizuka and Koketsu, 1964) and when the potential energy profile is symmetrical about the midpoint of the membrane and the external electrical potential difference is less than the height of the energy barrier (Hall *et al.*, 1973). An expression having a similar form has been derived by Jacquez (1971) for an arbitrary, continuous potential profile.
7. Integrated forms of the Nernst–Planck equation assume that the energy barriers within the membrane that must be overcome by the diffusing particle form a "continuum." These integrations are not generally valid when there are discrete energy barriers which result in discontinuities in the electrochemical potential of the diffusing species i along the diffusion pathway; hence the term *discontinuous diffusion* (Hille, 1975).

Chapter 3

1. The assumption that "bulk" or "macroscopic" electroneutrality is strictly preserved in solutions is central to all of our subsequent considerations. This, so-called *Nernst–Planck electroneutrality condition* can be expressed formally as follows:

$$\sum_c z_+ c_+ = -\sum_a z_- c_-$$

The (overwhelming) justification for this assumption is discussed by Guggenheim (1950) and Sten-Knudsen (1978). In essence it can be shown that minute departures from electroneutrality would result in enormous changes in electric fields many times greater than those encountered in most electrodiffusional systems. For example, as discussed by Sten-Knudsen (1978, p. 64), if the anion and cation concentrations of a solution differed by 1 mEq/liter, the gradient of the electric field in that solution ($d^2\psi/dx^2$) would be approximately 10^{14} V/m^2. The electrical potential differences encountered in most electrodiffusional systems are less than 1 V and changes in the electric field are not detectable. Thus, the electroneutrality condition is an eminently reasonable assumption.

2. Formally, the assumption is that $\beta_i = \beta_+ = \beta_-$ and that β_i is independent of concentration. This assumption infers that the membrane is "neutral" and does not discriminate between the anion (−) and the cation (+). This assumption is not valid for "charged membranes," which discriminate between anions and cations; a formal treatment for the diffusion of a single salt across charged membranes has been derived by Teorell (1953) (see Section 3.5).
3. It is a relatively simple matter to extend this derivation to the case where the anions and cations are multivalent and the valences have the same absolute value; that is, $z_+ = |z_-|$.
4. Jacquez and Schultz (1974) have derived a general expression for the membrane potential of symmetrical cells under steady-state conditions which makes no assumptions regarding electrical potential profiles, partition coefficients, and so forth, and is valid in the presence of "rheogenic pumps" (see Section 6.6). This expression is a generalization of one derived earlier by Mullins and Noda (1963).
5. There is some confusion in the literature over the terms *transference number* and *transport number.* MacInnes (1961, p. 59) and Robinson and Stokes (1959, p. 44) do not distinguish between the two, whereas Helfferich (1962, p. 326) does. Strictly speaking, t_i is defined as the fraction of the total current carried by the *i*th ion under conditions of uniform concentration driven by $\Delta\psi$ alone (i.e., in the absence of interactions between *i* and the flow of solvent or other solutes). However, as discussed by Kirkwood (1954), it is quite legitimate to refer to the *overall* t_i of a membrane that separates two solutions having different concentrations of *i*. Under these conditions t_i is a complex function of the distributions of all diffusible ionic species in the membrane and the compositions of the surrounding solutions [see equation (3.19)]; nonetheless, it can be defined empirically for any given set of conditions.
6. For a rigorous derivation, see MacInnes (1961, pp. 461–65).
7. Obviously, all membranes are electrically neutral! The terms "neutral" and "charged" simply distinguish between the manner by which electroneutrality is assured. In a "neutral" membrane, the *structural* components are either uncharged or bear an equivalent number of oppositely charged groups; thus, electroneutrality is maintained by fixed or unvarying elements of the membrane. On the other hand, a "charged" membrane possess structural components that are capable of dissociating into a preponderance of *fixed* positive or *fixed* negative charges, and electroneutrality is maintained by the presence of free, *unfixed* counterions. The nature of these counterions will vary depending upon the compositions of the surrounding solutions (see Lakshminarayanaiah, 1969).

Chapter 4

1. The mole fraction of *i* is the number of moles of *i* in a system divided by the total number of moles of all substances. In a mixture of perfect gases at total pressure *P*, the partial pressure of gas *i* is

$$P_i = c_i RT = x_i P$$

(Glasstone, 1956). Thus, using the perfect gas analogy, once again, x_i is proportional to c_i and either term can be used in writing expressions for μ_i.

2. See Section 7.3 for a discussion of electrode reactions and MacInnes (1961) or Lakshminarayanaiah (1969) for discussions of electrochemical cells and reversible electrodes.
3. See discussion in Section 7.3.
4. Recall that $c_i^i > c_i^o$ so that $\Delta c_i < 0$ and J_i is negative.

Chapter 6

1. For general discussions of carrier-mediated transport and carrier-kinetics see Wilbrandt and Rosenberg (1965), LeFevre (1975), Jacquez (1964) and Stein (1967). An introduction to carrier-kinetics can be found in Neames and Richards (1972).
2. See Blumenthal and Katchalsky (1969) for a treatment of a system where the interaction between the carrier and the transported solute can take place within the membrane and a discussion of the effect of carrier association-dissociation rates on transport.
3. For more general treatments of cotransport, including cases where the carrier complex is charged, see Stein (1967), Wilbrandt (1972), Heinz et al. (1972), Jacquez (1972), and Geck and Heinz (1976).
4. The approach we will take in this section follows that of Rosenberg and Wilbrandt (1955).
5. More general treatments of the kinetics of rheogenic carrier-mediated transport are given by Läuger and Stark (1970), Hladky (1972), Hall et al. (1973), and Stein (1977). An analysis of the effect of electrical potential differences on ion-linked cotransport is given by Geck and Heinz (1976) and Jacquez (1972).
6. It should be emphasized that whereas the carrier model is perhaps the simplest explanation for "uphill" countertransport and cotransport, positive (cotransport) as well as negative (countertransport) coupling between tracer flows can, in principle, result from membrane heterogeneities and need not involve "carriers" (Kedem and Essig, 1965; Patlak and Rapoport, 1971; Li and Essig, 1976, 1977).

Chapter 7

1. For an elegant and scintillating discussion of the application of electrical circuit models to biological membranes, see Cole (1968).
2. Clearly, when $\Delta\psi = 0$ and $c_i^o = c_i^i$ the unidirectional fluxes across the membrane are simply $J_i^{oi} = J_i^{oi} = P_i c_i$. Thus, under these ("short-circuit") conditions $G_i = J_i^{oi}(z_i^2 \mathcal{F}^2/RT)$. When $z_i = \pm 1$, J_i^{oi} expressed in microequivalents per centimeter squared hour is almost numerically equal to G_i expressed in millimhos per centimeter squared.
3. From equation (7.4) it follows that the integral conductances are

$$G_i = \frac{1}{\int_o^i (dx/u_i z_i^2 \mathcal{F}^2 \bar{c}_i)}$$

and

$$G_m = \sum_{i}^{n} \left[\frac{1}{\int_o^i (dx/u_i z_i^2 \mathcal{F}^2 \bar{c}_i)} \right]$$

Thus, in general, the overall t_i is a function of the concentrations and distributions of all i in the membrane which, in turn, are influenced by the concentrations in the surrounding solutions and the partition coefficients (see Section 3.3). Clearly, when both solutions have identical compositions

$$\left(\frac{G_i}{G_m}\right) = t_i = \frac{u_i z_i^2 \beta_i c_i}{\sum_i u_i z_i^2 \beta_i c_i}$$

which is identical with equation (3.19).

4. The direction of an electrical current is defined as positive when it flows from the anode to the cathode. Thus, strictly speaking, the I_{sc} is in the *same* direction as the active flow of i and, in fact, *is the flow of i* across the membrane. Accordingly, in essence, when $\Delta\psi = 0$, the active transport of *i draws* an equivalent amount of current from the battery through the external circuit to complete the internal circuit. However, I have found that it is difficult for "beginners" to appreciate this concept whereas they readily accept the notion that the I_{sc} "bucks out" the current resulting from ${}_aJ_i$. This notion is contrary to the conventional definition of the direction of current flow but is conceptually equally valid.

References

Andersen, B., and H. H. Ussing, *Acta Physiol. Scand.* 39:228, 1957.

Angrist, S. W., and L. G. Hepler, *Order and Chaos,* Basic Books, New York, 1967.

Armstrong, C. M., *Biophys. J.* 15:932, 1975.

Arndt, R. A., and L. D. Roper, *Simple Membrane Electrodiffusion Theory,* Physical Biological Sciences Miscellaneous, Blacksburg, Virginia, 1972.

Barr, L., *J. Theoret. Biol.* 9:351, 1965.

Beck, R. E., and J. S. Schultz, *Science* 170:1302, 1970.

Behn, U. A. R., *Ann. Physik* 62:54, 1897.

Biber, T. U. L., and P. F. Curran, *J. Gen. Physiol.* 51:606, 1968.

Blumenthal, R., S. R. Caplan, and O. Kedem, *Biophys. J.* 7:735, 1967.

and A. Katchalsky, *Biochim. Biophys. Acta* 173:357, 1969.

Caplan, S. R., and A. Essig, in *Current Topics in Membranes and Transport,* Vol. 9 (F. Bronner and A. Kleinzeller, eds.), Academic, New York, 1977, p. 145.

Cole, K. S. *Membranes, Ions and Impulses,* University of California Press, Berkeley, 1968.

Conti, F., and G. Eisenman, *Biophys. J.* 5:247, 1965.

Biophys. J. 6:227, 1966.

Curran, P. F., and S. G. Schultz, in *Handbook of Physiology,* Section 6, The Alimentary Canal (C. F. Code, ed.), American Physiology Society, Washington, D. C., 1968, p. 1217.

Dawson, D. C., *J. Membrane Biol.* 31:351, 1977.

DeDonder, T., and P. van Rysselberghe, *Thermodynamic Theory of Affinity,* Stanford University Press, Stanford, 1936.

De Groot, S. R., *Thermodynamics of Irreversible Processes*, North-Holland, Amsterdam, 1952.

and P. Mazur, *Non-Equilibrium Thermodynamics*, North-Holland, Amsterdam, 1962.

Denbigh, K., *The Principles of Chemical Equilibrium,* Cambridge University Press, Cambridge, Engl., 1966.

Dunlop, P. J., *J. Phys. Chem.* 61:994, 1957.

and L. J. Gosting, *J. Phys. Chem.* 63:86, 1959.

Durbin, R. P., *J. Gen. Physiol.* 44:315, 1960.

Eisenman, G. (ed.) *Glass Electrodes for Hydrogen and Other Cations,* Marcel Dekker, New York, 1967.

(ed.) *Membranes,* Volume II, Marcel Dekker, New York, 1973.

Essig, A., *J. Theoret. Biol.* 13:63, 1966.

O. Kedem, and T. L. Hill, *J. Theoret. Biol.* 13:72, 1966.
and S. R. Caplan, *J. Gen. Physiol.* 8:1434, 1968.
Eyring, H., R. Lumry, and J. W. Woodbury, *Rec. Chem. Prog.* 10:100, 1949.
Faxen, H., *Arkiv. Matematik, Astronomi. Fysik* 17:1, 1923.
Ferry, J. D., *J. Gen. Physiol.* 20:95, 1936.
Finkelstein, A., and A. Mauro, *Biophys. J.* 3:215, 1963.
Franz, T. J., and J. T. Van Bruggen, *J. Gen. Physiol.* 50:933, 1967.
Fujita, H., and L. J. Gosting, *J. Phys. Chem.* 64:1256, 1960.
Geck, P., and E. Heinz, *Biochim. Biophys. Acta,* 443:49, 1976.
Glasstone, S., *Thermodynamics for Chemists*, Van Nostrand, New York, 1947.
K. J. Laidler, and H. Eyring, *The Theory of Rate Processes,* McGraw-Hill, New York, 1941.
Goldman, D. E., *J. Gen. Physiol.* 27:37, 1943.
Grim, E., and K. Sollner, *J. Gen. Physiol.* 40:887, 1957.
J. Gen. Physiol. 44:381, 1960.
Guggenheim, E. A., *Thermodynamics*, North-Holland, Amsterdam, 1950.
Gunn, R. B., and P. F. Curran, *Biophys. J.* 11:559, 1971.
Hall, J. E., C. A. Mead, and G. Szabo, *J. Membrane Biol.* 11:75, 1973.
Heckmann, K., *Biomembranes* 3:127, 1972.
Heinz, E., P. Geck, and W. Wilbrandt, *Biochim. Biophys. Acta* 255:442, 1972.
Helfferich, F., *Ion Exchange,* New York, McGraw-Hill, 1962.
Henderson, P., *Z. Physik. Chem.* 59:118, 1907.
Z. Physik. Chem. 63:325, 1908.
Hill, T. L., and O. Kedem, *J. Theoret. Biol.* 10:399, 1966.
Hille, B., in *Membranes–A Series of Advances* (G. Eisenman, ed.), Marcel Dekker, New York, 1975, p. 255.
and W. Schwarz, *J. Gen. Physiol.* 72:409, 1978.
Hladky, S. B., *Bull. Math. Biophys.* 27:79, 1965.
J. Membrane Biol. 10:67, 1972.
Drugs and Transport Processes, University Park Press, Baltimore, 1974, p.193.
Hodgkin, A. L., and B. Katz, *J. Physiol.* (*London*) 108:37, 1949.
and R. D. Keynes, *J. Physiol. (London)* 128:61, 1955.
Hoshiko, T., and B. D. Lindley, *J. Gen. Physiol.* 50:729, 1967.
House, C. R., *Water Transport in Cells and Tissues,* Edward Arnold, London, 1974.
Jacquez, J. A., *Biochim. Biophys. Acta* 79:318, 1964.
Math. Biosci. 12:185, 1971.
Math. Biosci. 13:71, 1972.
and S. G. Schultz, *Math. Biosci.* 20:19, 1974.
Johnson, K. R., *Ann. Physik* (*Leipzig*) 14:995, 1904.
Katchalsky, A., and P. F. Curran, *Nonequilibrium Thermodynamics in Biophysics,* Harvard University Press, Cambridge, Mass., 1965.
Kedem, O., in *Membrane Transport and Metabolism* (A. Kleinzeller and A. Kotyk, eds.), Czechoslovakian Academy of Science, Prague, p. 87, 1961.
and S. R. Caplan, *Trans. Faraday Soc.* 61:1897, 1965.
and A. Essig, *J. Gen. Physiol.* 48:1047, 1965.
and A. Katchalsky, *Biochim. Biophys. Acta* 27:229, 1958.
J. Gen. Physiol. 45:143, 1961.
Trans. Faraday Soc. 59:1918, Part 1, 1963*a*.

Trans. Faraday Soc. 59:1931, Part 2, 1963*b*.
Trans. Faraday Soc. 59:1941, Part 3, 1963*c*.
Kimizuka, H., and K. Koketsu, *J. Theoret. Biol.* 6:290, 1964.
Kirkwood, J. G., in *Ion Transport Across Membranes* (H. T. Clarke, ed.), Academic, New York, 1954, p. 119.
Lakshminarayanaiah, N., *Transport Phenomena in Membranes,* Academic, New York, 1969.
Läuger, P., *Science,* 178:24, 1972.
Biochim. Biophys. Acta 311:423, 1973.
and G. Stark, *Biochim. Biophys. Acta* 211:458, 1970.
LeFevre, P., in *Current Topics in Membranes and Transport*, Vol. 7 (F. Bronner and A. Kleinzeller, eds.), Academic, New York, 1975, p.109.
Levi, H., and H. H. Ussing, *Acta Physiol. Scand.* 16:232, 1948.
Lewis, G. N., and M. Randall, *Thermodynamics,* 2nd edition, McGraw-Hill, New York, 1961.
Li, J. H., and A. Essig, *J. Membrane Biol.* 29:255, 1976.
Biochim. Biophys. Acta 465:421, 1977.
Lundegårdh, H., *Naturwissenschaften* 23:313, 1935.
MacGillivray, A. D., and D. Hare, *J. Theoret. Biol.* 25:113, 1969.
MacInnes, D. A., *The Principles of Electrochemistry,* Dover, New York, 1961.
Meyer, K. H., and J. F. Sievers, *Helv. Chim. Acta* 19:649, 1936.
Miller, D. G., *Chem. Rev.* 60:15, 1960.
Mitchell, P., *Adv. Enzymol.* 29:33, 1967.
Membranes and Ion Transport, Vol. 1, (Bittar, E. E., ed.) Wiley (Interscience), London, 1970, p. 192.
Mullins, L. J., *Ann. N.Y. Acad. Sci.* 94:390, 1961.
and K. Noda, *J. Gen. Physiol.* 47:117, 1963.
Neames, K. D., and T. G. Richards, *Elementary Kinetics of Membrane Carrier Transport,* John Wiley, New York, 1972.
Onsager, L., *Phys. Rev.* 37:405, 1931*a*.
Phys. Rev. 38:2265, 1931*b*.
Osterhout, W. J. V., *Ergeb. Physiol.* 35:967, 1933.
Proc. Nat. Acad. Sci. 21:125, 1935.
Pappenheimer, J. R., *Physiol. Revs.* 33:387, 1953.
E. M. Renkin, and L. M. Borrero, *Am. J. Physiol.* 167:13, 1951.
Parlin, R. B., and H. Eyring, *Ion Transport Across Membranes*, (Clark, H. T., ed.) Academic, New York, 1954, p. 103.
Patlak, C. S., and S. I. Rapoport, *J. Gen. Physiol.* 57:113, 1971.
Planck, M., *Ann. Physik. Chem.* 39:161, 1890*a*.
Ann. Physik. Chem. 40:561, 1890*b*.
Pleijel, H., *Z. Physik. Chem.* 72:1, 1910.
Prigogine, I., *Thermodynamics of Irreversible Processes,* John Wiley, New York, 1961.
and R. Lefever, *Membranes, Dissipative Structures and Evolution* (G. Nicolis and R. Lefever, eds.), John Wiley, New York, 1975, p. 1.
Renkin, E. M., *J. Gen. Physiol.* 38:225, 1954.
Robinson, R. A., and R. H. Stokes, *Electrolyte Solutions*, 2nd edition, Academic, New York, 1959.
Rosenberg, T., *Acta Chem. Scand.* 2:14, 1948.
and W. Wilbrandt, *Exp. Cell Res.* 9:49, 1955.
J. Gen. Physiol. 41:289, 1957.
Sandblom, J. P., and G. Eisenman, *Biophys. J.* 7:217, 1967.
G. Eisenman, and E. Neher, *J. Membrane Biol.* 31:383, 1977.

Scatchard, G., *J. Am. Chem. Soc.* 75:2883, 1953.
Schlögl, R., *Z. Physik. Chem.* 1:305, 1954.
Fortschritte der physikalischen Chemie: Band 9, 1964.
Schultz, S. G., in *Biological Membranes* (R. Dowben, ed.), Little, Brown, Boston, 1968, p. 59.
Fed. Proc. 38:2024, 1979.
and P. F. Curran, *Physiol Rev.* 50:637, 1970.
R. A. Frizzell, and H. N. Nellans, *J. Theoret. Biol.* 65:215, 1977.
Schwartz, T. L., *Biophys. J.* 11:596, 1971*a*.
Biophys. J. 11:944, 1971*b*.
in *Biophysics and Physiology of Excitable Membranes* (W. J. Adelman, Jr., ed.), Van Nostrand Reinhold, New York, 1971*c*. p. 47.
Simon, W., and W. E. Morf, in *Membranes,* Vol. 2 (G. Eisenman, ed.), Marcel Dekker, New York, 1973, p. 329.
Sjodin, R. A., *J. Gen. Physiol.* 48:777, 1965.
Staverman, A. J., *Rec. Trav. Chim.* 70:344, 1951.
Trans. Farady Soc. 48:176, 1952.
Stein, W. D., *The Movement of Molecules Across Cell Membranes,* Academic, New York 1967.
Biochim. Biophys. Acta 467:376, 1977.
Sten-Knudsen, O., in *Membrane Transport in Biology,* Vol. 1 (G. Giebisch, D. C. Tosteson, and H. H. Ussing, eds.), Springer-Verlag, Berlin, 1978, p. 5.
Talen, J. L., and A. J. Staverman, *Trans. Faraday Soc.* 61:2800, 1965.
Teorell, T. *Proc. Soc. Exp. Biol. Med.* 33:282, 1935.
Arch. Sci. Physiol. 3:205, 1949.
Prog. Biophys. Chem. 3:305, 1953.
Urban, B. W., S. B. Hladky, and D. A. Haydon, *Fed. Proc.* 37:2628, 1978.
Ussing, H. H. *Acta Physiol. Scand.* 19:43, 1949.
Adv. Endocrinology 13:21, 1952.
Ann. N.Y. Acad. Sci. 137:543, 1966.
in *Perspectives in Membrane Biophysics* (D. P. Agin, ed.), Gordon and Breach, New York, 1972.
and K. Zerahn, *Acta Physiol. Scand.* 23:110, 1951.
Widdas, W. F., *J. Physiol.* (*London*) 118:23, 1952.
Wilbrandt, W., *J. Membrane Biol.* 10:357, 1972.
and T. Rosenberg, *Pharmacol. Rev.* 13:109, 1961.

Frequently used symbols

A affinity of a chemical reaction (calories/mole)
c_i concentration of species i in bulk solution (see Figure 2.1)
$\overline{c}_i$ concentration of species i in the membrane (see Figure 2.1)
$\dot{c}_i$ mean concentration of species i
D_i diffusion coefficient of species i (cm^2/sec)
E total electromotive force (volts)
E_i electromotive force of species i (volts)
$\mathcal{F}$ the Faraday (96,500 coulombs/equivalent)
G Gibbs free energy (calories)
G_i partial ionic conductance of species i (mhos/cm^2)
G_m total membrane conductance (mhos/cm^2)
I total current (A/cm^2)
I_i current attributable to the flow of i (A/cm^2)
J_i flux of species i (moles/cm^2 sec) (see Figure 2.1)
J_v rate of volume flow (cm/sec)
L_p hydraulic conductivity (filtration coefficient) (cm^3/dyne sec)
P pressure (dynes/cm^2; atm)
P_i permeability coefficient of membrane to species i (cm/sec)
q degree of coupling
R universal gas constant [8.13 J/(K mole), or 1.99 cal/(K mole), or 0.082 liter-atm/(K mole)]
R_i resistance to species i (ohms cm^2)
R_m total membrane resistance (ohms cm^2)
T absolute temperature (K)
t time (sec)
u_i mobility of species i [moles cm^2/(cal sec)]
V volume (cm^3)
$\overline{v}_i$ partial molal volume of species i (cm^3/mole)
x thickness of membrane (cm)
z_i valence of species i

β_i partition coefficient of species i
μ_i chemical potential of species i (cal/mole)
$\tilde{\mu}_i$ electrochemical potential of species i (cal/mole)
π osmotic pressure
σ_i reflection coefficient of species i
ψ electrical potential (volts) (see Figure 2.1)

Index